The Use of Gender Markers in Animals

DUISBURGER ARBEITEN ZUR SPRACH- UND KULTURWISSENSCHAFT

DUISBURG PAPERS ON RESEARCH IN LANGUAGE AND CULTURE

Herausgegeben von /edited by
Ulrich Ammon, René Dirven und /and Martin Pütz

Band/Volume 112

Zu Qualitätssicherung und Peer Review der vorliegenden Publikation

Die Qualität der in dieser Reihe erscheinenden Arbeiten wird vor der Publikation durch einen Herausgeber der Reihe geprüft.

Notes on the quality assurance and peer review of this publication

Prior to publication, the quality of the work published in this series is reviewed by one of the editors of the series.

Ludmila Zemková

The Use of Gender Markers in Animals

As Demonstrated by Issues of National Geographic

PETER LANG
EDITION

Bibliographic Information published by the Deutsche Nationalbibliothek
The Deutsche Nationalbibliothek lists this publication in the Deutsche Nationalbibliografie; detailed bibliographic data is available in the internet at http://dnb.d-nb.de.

ISSN 0934-3709
ISBN 978-3-631-67457-4 (Print)
E-ISBN 978-3-653-06623-4 (E-Book)
DOI 10.3726/978-3-653-06623-4

Peter Lang Edition is an Imprint of Peter Lang GmbH.

Peter Lang – Frankfurt am Main · Bern · Bruxelles · New York · Oxford · Warszawa · Wien

This publication has been peer reviewed.

www.peterlang.com

Contents

1. Introduction

"Gender is the most puzzling of the grammatical categories. It is a topic which interests non-linguists as well as linguists and it becomes more fascinating the more it is investigated" (Corbett 1991: 1).

The thesis aims to be a contribution to gender studies. Out of the many potential research challenges in this area, I have chosen to explore the problem of assigning gender to animal nouns.

The introductory part of the thesis will discuss the origin and history of gender to set the scene for more practically oriented parts. Chapter 2 will establish the theoretical background information for my further research.

In the following part the focus is predominantly on gender in animals. Subchapter 3.1. ("Gender in present-day English") summarizes recent descriptions of gender assignment in animal nouns. Subchapter 3.2. will look at grammarians' treatment of "gender in present-day English". Subchapter 3.3. describes basic types of assigning gender to animals in the world's languages. Detailed analyses of gender assignment in the corpora of this study will be at the core of Chapter 4.

2. Theoretical background

To avoid possible misunderstanding, the term of gender must be initially defined, because it is used to refer to a variety of concepts in linguistic literature. When the research focus is on gender as manifested in language, the usual objectives are to discuss the differences between male and female use of language, or the differences in the representation of women and men in a language.

2.1 The History of Gender

This chapter attempts to give a short overview of the history of gender, both generally and in comparison with selected languages. In this context, the following questions need to be answered:

- What is gender?
- What is the history and development of gender systems?
- What is the position of gender in the language system?

As Baron points out (1986: 91): "The history of gender has never been satisfactorily explained."

Studies of gender as a linguistic category have a long history. To understand the meaning of the linguistic term "gender", Hockett's definition could be a good starting point: "Genders are classes of nouns reflected in the behaviour of associated words" (Hockett 1958: 231). Saying that means that the determining criterion of gender is agreement. A language may have two or more classes of gender. The classification frequently corresponds to a real- world distinction of sex, at least in part, but often too it does not (Corbett 1991: 1).

The word gender is derived from Latin "*genus*" via Old French "*gendre*", meaning "class", "kind" or "sort" and it referred to the division of Greek nouns into three classes: masculine, feminine and neuter. These classes have nothing to do with the meaning of the word. They just label the formal properties. The word "gender" is used either for a group of nouns or for the whole category. The first theories of grammatical gender rely heavily on the connection between grammar and physiology. Many early grammarians noticed that gender may be a function of the form of a word rather than its meaning (Corbet 1991: 92). According to Aristotle "the first of the Greek grammarians to isolate the category of gender was Protagoras, who classified names as masculine, feminine and inanimate according to the sex, or lack of sex, of their referents" (cited in Baron 1986: 92). Aristotle himself classified nouns according to their terminations rather than referential meaning. (ibid.: 92). Masculine gender and men's language were considered as primary of creation and in importance, "the most important" (Goold Brown 1851- cited in Baron 1986: 3) and "the superior and more excellent" (James Harris 1751 cited in Baron 1986: 3). As Spender (1980: 3) points out, treating the masculine as the linguistic norm has been called "one of the most pervasive and pernicious rules that has been encoded".

Even today, many linguists assume that the masculine is "the normal, or unmarked" gender and that all English nouns are masculine unless specially marked.(ibid.: 97)

2.2 An Overview of Gender in English

2.2.1 History

English inherited a formal gender system from Germanic three-way grammatical classification. Between 10th and 14th centuries it was replaced by the semantic "natural" or "logical" gender system. This gradual change was completed in Early Middle English (cf. Dekeyser 1980: 102). The phonetic changes in the syllable led to the loss of most inflectional endings of the noun. It is probably not correct to classify the Old English gender system as a purely formal category, because "many suffixes were affiliated to more than one gender" (Kastovsky 2000: 712). The distribution of nouns into the three genders had a semantic basis in Old English, thus most male nouns were masculine, most female nouns feminine and the majority of neuter nouns were "asexual" (Jones 1988: 35). Moore (1921: 91) claims that "natural gender did not replace grammatical gender in Middle English but survived it". On the other hand, scholars like Mitchell (1985: 29) do not consider this analysis representative of the respective periods.

According to Curzan (2000), Latin grammar, its terminology and classification was the core of early English grammars. Feminine old English noun endings were generally more distinctive than masculine or neuter. The feminine gender thus lingered longer than the other two, and this continued association led to the persistence of feminine pronominal reference with some nouns to the present day (Fennell 2001: 64). The concrete developments of dative-accusative levelling in English is treated in Visser (1963: 427) and Howe (1996: 114).

The conclusion from these observations is resumed by Wagner in her dissertation thesis (2002: 41):

> The only conclusion to be drawn from those observations is that there was a rather extended period of time in the history of the English language when the choice of a

supposedly masculine personal pronoun (*him*) said nothing about the gender or sex of the referent. It could be masculine, male, neuter, or asexual – and every combination of those three.

Table 2.1: The distribution of nouns into the genders in Wagner (2002)

gender / sex	masculine	neuter
male	x	x
asexual	x	x

Although the English gender system is based on semantic criteria, where the meaning of a noun determines its gender and gender is reflected only in personal, possessive and reflexive pronouns, it is not totally dependent on the straightforward criteria of humanness and biological sex but may also be affected by pragmatic factors. For centuries, it has been confused with the biological category of sex.

Fortunately most English grammars did not worry their readers too much when it came to gender. To mention just a few grammarians, Anderson (cited in Baron: 96) offers thirteen gender distinctions. He also refines the neuter category to include inanimates as well as "animals that have no sex at all, those whose sex is not apparent, and others still in which, though the sex be known, it is not at all considered".

Most books in the eighteenth and nineteenth centuries did not discuss pronoun gender to any great extent. A good example is the work of Murray. He says simply that "gender is the distinction of sex" and describes four principal methods of distinguishing sex:

- By different words (*goose, cock, hen*)
- By a difference of termination (*duke, duchess*)
- By adding an adjective or pronoun to the substantive (*he-goat, she-goat*)
- By prefixing another substantive to the word (*cock-sparrow, hen-sparrow*) (ibid.: 97).

Grammarians treated the masculine gender as primary both in the natural world and in the sentence. They also continued to insist that personifications reflected particular sexual qualities: the horse is classified as masculine because of its usefulness and its generous nature, while the hare is feminine because of its timidity. The elephant is masculine because of its great strength and size, as well as its sagacity, fortitude and docility, and the cat is simply female, without any comment (Crombie 1830, cited in Baron 1986: 103).

Joly defines males as major powers, and so any animal is treated as masculine when it is "presented as an active power and possible danger", while the feminine pronoun, representative of minor power, is employed when the animal "is regarded as a potential prey". Generally, ***she*** will be said to any animal, big or small, that is in some way subordinated to the speaker (1975: 271).

In this respect, the most frequently issues discussed are those connected with the use of the generic masculine ***he, his, him*** and the use of ***generic man***. They evoke predominantly images of males.

At the end of the eighteenth century, prescriptive grammarians, such as Joseph Priestley, Hugh Blair, Lindley Murray started discussion on ***he*** or ***she*** and the ***singular they***. The singular ***they*** was used in written and spoken English prior to the nineteenth century.

"The first calls for the development of a new pronoun arose out of a concern not for the unfair exclusion of women from many English sentences but for the stylistic accuracy" (Baron 1986: 8). Prescriptive grammarians were the first who called for change.

The discussion continued with the feminist movement insistence that ***he*** should not be used when the referent includes women. The following three books, published in the mid 1970's: Lakoff "Language and Women's place", Key (1975) "Male/Female Language, and Thorne et al. (1983) "Language and sex: Difference and Dominance" have been accepted as milestones in gender and language studies.

Materials and books by authors like Tannen (1994), Cameron (1994), Pauwels (1998), Spender (1980) or Coates (1993) are primarily written from a feminist point of view.

Translation studies and the influence of feminism on translation theory and practice is described by Simon (1996). She points out that "translators and women have historically been the weaker figures in their respective hierarchies" (Simon 1996: 1).

To mention just a few authors dealing with the question of sexism in textbooks and EFL materials, Bodine (1975) offers both diachronic and synchronic descriptions of the issue. Poreca (1984) has conducted similar research. Research on British EFL materials is described in the work of Sunderland (1994).

3. Gender Assignment in Animal Nouns

3.1 Gender in Present-Day English

As stated above, the aim of the thesis is not to study the history of gender and feminist linguistic theories. In the following part I will focus my attention on assigning gender to animal nouns in present-day English.

To mention just a few works dealing with gender assignment in Modern English, modern views on the existence of the category of gender can be found in Vachek (1964), whose criteria for assigning gender are in agreement with the factors that are identified as crucial in my own analysis. He notes that "the shift of gender" is motivated by affect or emotion (Vachek 1964: 190). A special case is constituted by pronominal reference to nouns denoting small children and animals. When the sex of the individual concerned is either unknown or irrelevant (e.g. *fly*, *cat*, etc.), neuter reference is commonly used. On the other hand, feminine or masculine reference signals "some interest of the speaker in the individual referred to" (P.A. Erades 1956: 7, cited in Vachek), but the gender chosen need not necessarily tally with the actual sex of the individual: "there may be, again, some interference of emotional factors" (ibid.: 191). The author also mentions "some interference of emotive factors" and personifications which may often be effected on traditional, conventional lines. "These conventions can be squared with the tendencies governing pronominal reference to nouns denoting inanimate things: bigger and stronger animals rank most frequently as masculines, while smaller, gentler animals (especially singing birds) prevalently rank as feminines" (ibid.: 191).

Crystal (1995: 209) solves the problem of personification and the reasons for the prevalence of the feminine pronouns in "neuter" contexts. As he notes "Many nouns are given variable gender, depending on whether they are thought of in an intimate way. Pets are often *he* or *she*."

Brinton (2000-cited in Wagner 2002: 48) notes that "gender is generally a covert category" in nouns, while "a related category of animacy" is expressed in personal, interrogative and relative pronouns. Very interesting is the animacy grouping. She distinguishes two main groups- humans and higher animals in one group and lower animals and inanimates in the second. Animals thus emerge on both sides of the scale, as can be seen in the following scheme:

Figure 3.1: Gender categories in Brinton (cited in Wagner 2002: 48).

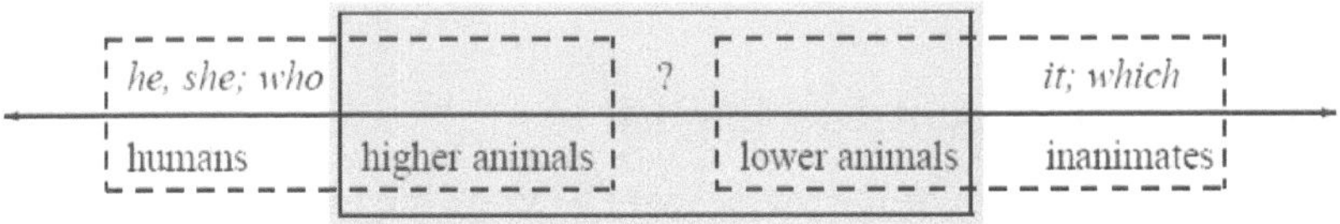

According to this scale it is clear that gender assignment depends on a wide variety of influential factors.

MacKay and Konishi (1980 cited in Wagner 2002: 121) investigated the use of "human" pronouns to refer to non-human antecedents in children's literature. One class contained animals (including real, imaginary, and toy animals). The major result from the counts was highly unexpected – 82% of pronominal references concerning animals were masculine (62%), and feminine (20%), respectively (cf. 2002: 122).

A similar research is described in Prčíková's doctoral thesis (Prčíková 1999). The corpus for her analysis includes children's stories published in Britain and the U.S. Seventy nine per cent of observed pronominal pronouns referring to animals were masculine, 10% feminine (cf.: 51).

Vomlela (2002) studied gender and animal nouns on selected web pages devoted to animal keeping and breeding. Concluding from the material analysed, most of the animals (predominantly *rabbit*, *reptile*, *dog* and *lizard* species) are masculine. *Cat*, *bird* and *snake* species are mainly feminine (cf. 2002: 140).

The gender of animals in modern Canadian English is investigated by Morris in her doctoral thesis (Morris 1991, cited in Wagner 2002). In her data, based on evidence and observations, "animals are much more frequently *he* than *she*" (cf. 140). The following scheme shows the hierarchical system of assigning gender to animals based on her research (Morris 1991 cited in Wagner 2002: 141):

Figure 3.2: Hierarchical system of assigning gender to animals (Morris 1991 cited in Wagner 2002: 141).

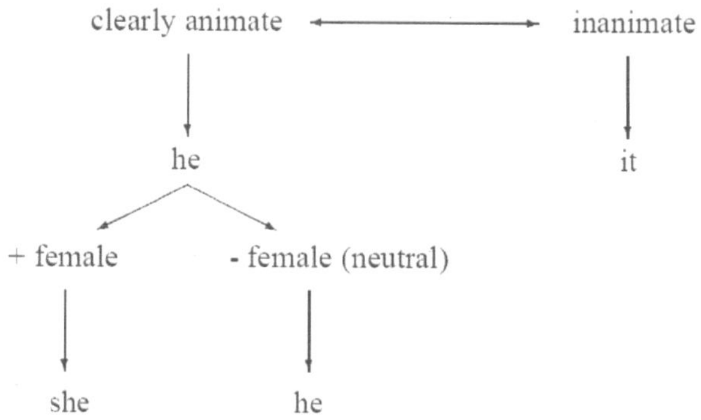

In Marcoux (1973) the students' use of personal pronouns in tag questions was investigated. He found a surprisingly high occurrence of "human" pronouns used to refer to animals of unknown sex. The masculine pronouns prevailed (Marcoux 1973: 104).

Hrnčíř (1997) summarizes all the three main markers of gender distinction which are signalled:

(a) By the suffix –ess, e.g.: *lion- lioness, tiger- tigress, leopard- leopardess*
(b) By different words, e.g.: *boar- sow, stallion- horse, cock- hen.*
The author pays attention to youngsters, e.g.: *foal- colt- filly, calf- bullock- heifer.*
(c) By compound nouns which contain words such as male- female (*frog- male frog- female frog- tadpole*), billy- nanny (*billy-goat,nanny-goat,kid*), tom- tabby (*tom-cat, tabby, catling*), he- she (*he- wolf, she- wolf, wolf cub*), dog- bitch (*dog-ferret, bitch- ferret, kit*), bull- cow (*bull moose, cow moose, moose calf*), buck- doe (*kangaroo-buck, doe, kid*).

The author also mentions gender distinctions which are typical of specific kinds of animals: e.g. goat: *billy-goat, nanny-goat, kid.*

cat: *tomcat, pussy-cat, puss, tabby, kitten, kit, kitty, catling.*

"Lower animals" are also mentioned, e.g. *squirrel, frog, ant.* The following table summarizes all three ways of gender distinction in "higher animals" and some "lower animals", e.g. rat: *buck – doe – nestling.*

Table 3.1: Gender distinction in animal nouns (Hrnčíř 1997: 166–167)

animal	male	female	youngster
antelope	buck	doe	kid, calf
ass, donkey	jack, jackass, dicky	jenny, she-ass	foal, colt, filly
badger	boar	sow	cub
bear	he- bear	She- bear	cub
boar	boar	sow	piglet
cat	tom- cat, tom, gib-cat	pussy-cat, puss, tabby, she- cat	kitten, kit, kitty, catling
cattle	bull, ox	cow	calf, heifer, bullock, steer
chicken, fowl(s)	cock, rooster	hen	chick
deer	buck, stag, hart	doe, hind	fawn, calf, kid
dog	dog, hound	bitch	pup, puppy, whelp
duck	drake	duck	duckling
elephant	bull	cow	calf
elk moose	bull	cow	calf
fallow deer	buck	doe	fawn
ferret	dog, buck, jack,hob	bitch, doe, jill	kit
fox	dog, dog-fox, vix	bitch, bitch-fox, vixen	cub, pup, whelp, kit
frog	male frog	female frog	tadpole
goat	billy, billy- goat, buck, he- goat	nanny, nanny- goat, doe, she- goat	kid
goose	gander	goose	gosling
hare	buck, jack	doe, puss	leveret
horse	stallion, horse, sire, stud, gelding	mare	foal, colt, filly
kangaroo	buck, boomer	doe, blue- flier	joey
leopard	leopard	leopardess	cub
lion	lion	lioness	cub, whelp
otter	dog-otter, dog	bitch-otter, bitch	cub, kid, pup
peafowl	peacock	peahen	chick
pheasant	cock, cock- pheasant	hen, hen-pheasant	chick, poult

animal	**male**	**female**	**youngster**
pig, hog	boar, hog,	sow, gilt	piglet, piggy, pigling
polecat	hob	gill	kit
rabbit	buck-rabbit, buck	doe-rabbit, doe	nestling, kit, kitten, bunny
rat	buck	doe	nestling
red deer	stag, hart	hind	fawn
reindeer	buck	doe	fawn
rhinoceros	bull	cow	calf
roe, roedeer, roe- deer	roebuck, roe- buck, buck	roedoe, roe- doe, doe	fawn
seal	bull	cow	pup, puppy, cub
sheep	Ram, tup, wether	ewe	Lamb, hog, teg
sparrow	cock- sparrow, cock	hen- sparrow, hen	chick
tiger	tiger	tigress	cub, whelp
turkey	turkey- cock, cock	turkey- hen, hen	poult
walrus	bull	cow	calf
whale	bull	cow	calf
whale	bull	cow	calf
wolf	he- wolf, dog	she- wolf, bitch	cub, whelp
zebra	stallion	mare	foal, colt, filly

The following conclusions (highly valuable for my further research) can be drawn from the above mentioned observations:

- **A number of emotive factors and some degree of personal involvement play a role in the choice of pronouns when referring to animals.**

My research has revealed that emotional involvement is crucial when assigning gender to animals. A good example can be the *panda* species, where feminine and masculine gender distinctions dominate over the neuter gender. The owners and researchers are very likely to refer to the panda as ***he*** or ***she***, signifying their emotional attitude or intimacy towards this endangered and critically threatened animal (as can be seen in subchapter 4.2.6. and in examples F.24.–F.43. included in the appendices)

- **The cut-off point within the class of animals differs from speaker (or author) to speaker, depending on their background, professions, etc.**

Someone who grew up in a big city will probably restrict the ***he*** **–** ***she*** distinction only to pets. For a lay person, a strange animal such as a *honey badger* or a *kinkajou* (which the speaker may not ever have seen) will be ***it***, whereas researchers and zoologists refer to a wide variety of strange animals as *he* or *she*, as demonstrated in Chapter 4.6.

- **When referring to animal nouns, neuter pronouns are the least expected forms, whereas the masculine dominates.**

The findings offered by the authors mentioned above have shown masculine to be the prevailing gender, whereas neuter has been the least frequent form. The analysis of the data collected in my research in *National Geographic* corpus reveals the prevalence of feminine, whereas the neuter proves to be a marginal. My contribution in this respect can be found in subchapter 4.2.8.

The following section will look at grammarians' treatment of gender in Present- Day English. As Vachek notes, "normative grammars keep to the well-known traditional rule stating that the nouns denoting male beings should be classed as masculines, those denoting female beings as feminines, and those referring to inanimate objects as neuters" (ibid.: 189). A look at modern reference grammars reveals that this traditional approach is stronger than common sense. Gender has always been described as a formal category of English, but far removed from a true gender language.

The following overview is focused on the category of animal nouns, which is crucial for my further analyses. Chapter 3.2. will summarize the grammarians' views on gender in present-day English.

3.2 An overview of gender in animals in modern reference grammars

Most grammars tell us that the appropriate pronoun to use when referring to an animal is *it*, except for cases where the sex of the animal is known. The following grammars will be under study:

- Dušková, L. (1988) *Mluvnice současné angličtiny na pozadí češtiny*. Praha: Academia
- Hais, K. (1975) *Anglická mluvnice*. Praha
- Huddleston, R. and Pullum, G.K. (2002) *The Cambridge Grammar of the English Language*. Cambridge: Cambridge University Press

- Biber, D., Johansson, S., Leech, G., Conrad, S. and E. Finegan (1999). *Longman Grammar of Spoken and Written English.* London: Longman.
- Quirk, R., Greenbaum, S., Leech, G., Svartvik, J. (1985). *A Comprehensive Grammar of the English Language.* Longman
- Greenbaum, S. and R. Quirk (1993) *A Student's Grammar of the English Language.* Longman
- Alexander, L.G. (1998) *Longman English Grammar.* Longman

The most comprehensive grammar book often used by Czech students "*Mluvnice současné angličtiny na pozadí češtiny*" (1988) by L. Dušková et.al. puts in contrast the categories of gender in Czech and English. The authors provide us with examples of how gender is expressed with animal nouns in English.

Dušková et al. stress the difference between the primarily grammatical category of gender in Czech with limited semantic base and the primarily semantic gender category in English. Most animal nouns do not distinguish the gender and are referred to by ***it***. The usage of this pronoun is typical of scientific texts and literature: "*The cat flung itself across the yard. The fish produced every trick to detach itself from its tormentor*" (1988: 86). But Dušková points out that "the use of he or she does not necessarily express the sex, e.g. a fisherman speaks about a big fish as *he*, about whales as *she*." According to Dušková, in these cases "the only relative pronoun used is *which*" (1988: 86). This is in contrast with English materials and grammar books, for example Greenbaum and Quirk (1993).

According to Dušková (1988), persons of masculine sex are masculine, persons of feminine sex are feminine and inanimates and animals are neuter. Higher animals stay somewhere between the persons and inanimates. Stylistic and emotive factors can cause exceptions from the basic classification.

A more general description of the categories of gender can be found in Anglická mluvnice (1975) by Karel Hais.

Hais (1975: 62) notes that animal names do not distinguish the gender and they are referred to by ***it.*** Only domestic animals and pets that are useful to people are referred to according to their sex. More formal gender distinction appears in higher animals (*dog- bitch, cat- tomcat, bull- cow, stallion- mare, gander- goose, lion- lioness, he- goat- she- goat*, etc.) Lower animals are referred to by ***it*** (*snake*). According to Hais (1975: 62) the animals which "are in contact with human beings are of masculine gender (*dog, horse, canary*) or of feminine gender (*cat, parrot, fish*)."

Dušková (1988: 82–87) and Hais (1975: 59–62) distinguish two main groups:

- The first group is formed by pairs expressed by different lexical items, such as: *bull- cow, cock- hen, mare- colt- filly, drake- duck, fox- vixen*

- The second group is formed by the pairs of animals of opposite sex which are distinguished by word formative processes:
  - by suffixes: *lion- lioness, tiger- tigress*
  - by compound words: *wolf- she-wolf, he-goat and billy goat* (or *he- goat*)- *she- goat* (or *nanny- goat*), *turkey- cock- turkey- hen, tom- cat- pussy- cat*

As for English grammars, I will have a look at the following:

To my knowledge, the most comprehensive explanations concerning the gender of animals can be found in Huddleston and Pullum's *Cambridge Grammar of the English Language* (2002).

The authors distinguish the degree to which gender is "grammaticalised" in English (on the problem of "grammaticalization", cf. Hopper and Traugott 1993).

They contend (2002: 485) that "while pronoun choice depends on the meaning or reference of the antecedent there are places where the linguistic form of the antecedent restricts the choice of the pronoun." The author compares the following pair of examples:

a. The dog has lost his/its bone   b. *Fido has lost his/*its bone*

According to the author the male animal mentioned above could be referred to by both *The dog* and *Fido*, "but the fact that the latter has a proper noun as head excludes the use of neuter pronoun ***it*** that is found as an alternative to *he* in *a*." (ibid.: 485). He compares this situation to that with human babies.

Animals are treated as non-humans. The book provides us with these examples:

- *The bull turned his/its head.*
- *The cow was lying on her/its back.*
- *The dog looked as if he/she/it needed a good brush.*

The sentences above show that the meaning of it does not mean "neither male nor female". The authors claim that in many cases the neuter pronoun is used because the speaker does not know what the sex of the animal is.

According to the authors, the difference between ***it*** and ***he*** or ***she*** does not lie in the referent itself: "it is a matter of whether the speaker chooses to encode the sex of the referent. Encoding of sex in the antecedent does not require that it also be encoded in the pronoun." (ibid.: 489).

To sum up, the use of feminine and masculine in the case of animal antecedents is more likely with pets and domestic animals. Apart from distinguishing these two categories, Huddleston et al. mention creatures ranking high in the kingdom of wild animals (such as *lions, tigers, elephants*, etc.). They also mention the importance of "greater degree of interest in or empathy" with the animal (on the issue of socio-pragmatic gender studies, cf. Mathiot and Roberts 1979). The animal is thus referred to by ***he*** or ***she***. The use of masculine and feminine is even more obligatory if the animal is referred to by a proper name (ibid.: 489). As will be shown in Chapter 4, the analysis from *National Geographic* confirms this observation.

The authors also distinguish three groups of dual-gender nouns:

- Dual-gender masculine/neuter nouns, containing mainly farm animals such as *bull, cock, gander, he-goat, ram, stallion, tom-cat*
- Dual-gender feminine/neuter nouns. This group contains predominantly names for the female animals and female kinship terms connected with those in the first group.
  Cow, hen, lioness, mare, she-goat
- Triple-gender nouns represent the terms denoting animals without specification of sex.
  Blackbird, dog, elephant, frog, goat, horse, lion, octopus, snake

Masculine and feminine are generally less likely to be used for lower animals except pets.

Longman's Grammar of Spoken and Written English (Biber et al. 1999) offers an approach to gender in animals. The authors pay attention to spoken language and consider gender "a less important category in English than in many other languages" (Biber et al. 1999: 311)

The authors distinguish two main gender classes: personal and non-personal:

Figure 3.3: Gender classes in Biber (Biber et al. 1999: 311f).

	example noun	pronoun
personal/human:		
masculine	*Tom, a boy, the man*	*he*
feminine	*Sue, a girl, the woman*	*she*
dual	*a journalist, the doctor*	*he, she*
non-personal/neuter:	*a house, a bird*	*it*

Animals (pets in particular, *ibid.:* 318) fall into the category of a three-way choice (personal *he, she and non-personal it*).

Longman English Grammar (1988: 49–50) by L. G. Alexander offers very limited information on the gender of animals.

Alexander identifies the masculine and feminine of animal nouns by:

- "contrasting nouns describing animals (normally replaceable by "it")
  bull- cow, cock(or rooster)- hen, dog- bitch, gander- goose, pig- sow,ram- ewe, stallion- mare"
- "–ess endings and other forms indicating sex/ gender": *leopard- leopardess, lion- lioness, tiger- tigress*
- "– the stressed prefix *he-/ she- he- goat/ she- goat, he- wolf/ she- wolf"*.

Clearly arranged explanations concerning the gender of animals can be found in Quirk et al. (1985). The authors divide animal names into two categories: higher and lower animals.

- Higher animals
  Male/ female gender distinctions in animal nouns are maintained by people with special concern (for example with pets, eg: *cock* and *rooster* for the male (with *which- he/it* or *who- he* coreference and hen *for* female (with which-she/it or who-she coreference (Quirk et al. 1985: 317)
  Quirk *et al.* give other examples:
  Buck-doe, bull-cow, dog-bitch, gander- goose, lion- lioness, ram-ewe, stallion-mare, tiger-tigress.
  In general, in nonexpert contexts there is no need to make a gender distinction, such as *dog- bitch, stallion- mare.* We can then use one term to cover both sexes, as in the case of *dog* and *lion*, or use a different term, eg: *horse* to cover both *stallion and mare.*(ibid.: 317)
- Lower animals
  Lower animals (eg: *beetle, butterfly, snake, toad, tadpole*) have *which* and *it* as pronouns. However, lower animals may also be viewed as higher animals. Thus we may speak of '*goldfish who swim around*', '*bees who are busy*', etc.
  Sex differences can be indicated by a range of gender markers for any animate noun where they are felt to be relevant, eg: *male- female frog.*

The book provides us with the following examples: *buck-rabbit~ doe-rabbit, cock-pheasant~ hen- pheasant, dog-fox~bitch-fox, he-goat~ she-goat, roe-buck~ roe-doe.*

According to the authors, some animals require finer gender distinctions in language than others. This can be attributed to a number of factors. Laymen normally have no knowledge about the sex of animals like *ant, herring, snake, spider*; or, even if they do, it may not be a fact that they want or need to indicate. *He* and *she* are only likely to be used for animals with which man, "the speaking animal", has the closest connections- with the domesticated animals.

Quirk et al. distinguish nine gender classes. Three of them concern animal nouns (higher male animal, higher female animal and lower animal) as illustrated in the scheme below (ibid.: 314):

Figure 3.4: Gender system according to Quirk (Quirk et al. 1985: 314).

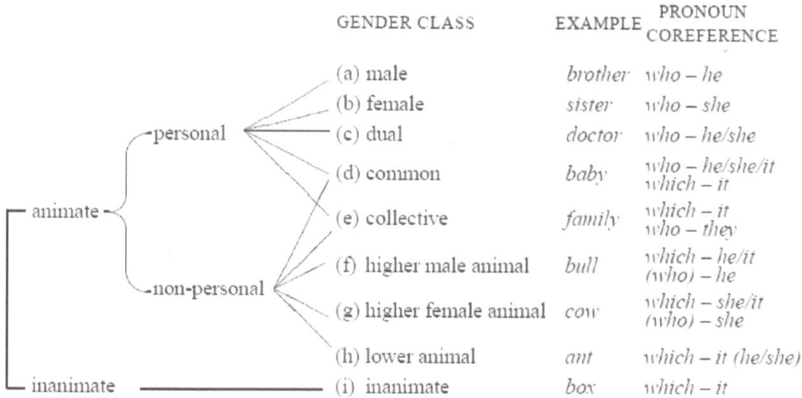

A very similar classification of nouns referring to animals is presented in Greenbaum, S. -R. Quirk: *A Student's Grammar of the English Language* (1993). The animals are divided into two main groups- familiar and less familiar animals. The following scheme distinguishes two main gender classes of animals:

Figure 3.5: Classification of animal nouns in Greenbaum et al. (1993: 100).

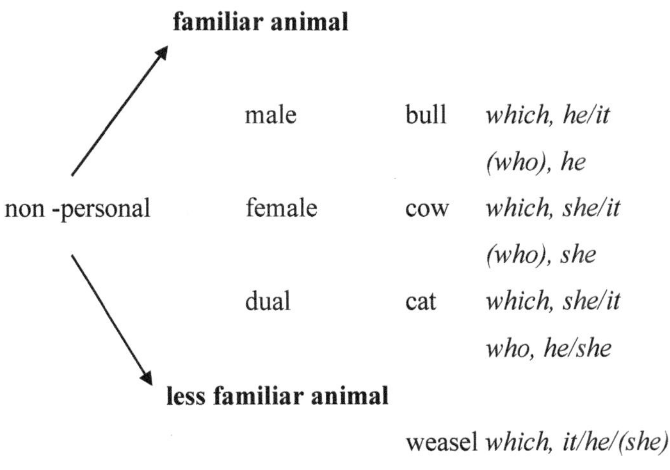

The above scheme differs from the one offered in Quirk et al. (1985) in that it introduces one additional category and it is dual gender of animals. In spite of the fact that there exist pairs like *bull* and *cow, ram* and *ewe, hen* and *cock, dog* and *bitch,* one member of the pair is used with dual gender.

Less familiar animals constitute the majority of animals, they are treated as inanimate and are referred to by *it* but they may be referred to as *he* or *she* as well.

The above mentioned grammars by Quirk et al. (1985) and by Greenbaum et al. (1993) pay attention mainly to the terminological differentiation and professional language use. No reference is made to the assignment of animal gender in spoken English.

Other comprehensive grammars and practise books as Quirk, R. Greenbaum, S., Leech, G. And Svartvik, J.: *A Grammar of Contemporary English* (Longman) (1972), Huddleston, R.: *Introduction to the Grammar of English* (CUP) (1984), Huddleston, R. *English Grammar: An outline* (CUP) (1988), Murphy: *English Grammar in Use,* CUP, (1990), Beaumont- Granger: *The Heinemann English Grammar,* (1989). Swan- Walter: *How English Works, A Grammar Practise book,* OUP (1997), Azar: *Understanding and Using English Grammar* (1989) do not bring more information concerning gender in animal names.

To sum up, in discussing gender of animals more attention in grammar books is given to the word-formative processes or lexical items distinguishing heteronymous animal pairs than to the ways of assigning gender to unmarked nouns. Most grammars, not very clearly, divide animals into familiar, less familiar or higher and lower animals. According to the majority of grammars animals are treated as inanimates and thus referred to by *it.* This gender system may look confusing to speakers of English. The authors pay attention mainly to terminological differentiation, professional language use and gender assignment, mainly in written English. Most authors do not mention spoken reality when referring to animals. Only in more comprehensive grammars other factors of animal gender assignment are mentioned, namely:

- **greater degree of interest in and empathy with the animal,**
- **its position in the kingdom of wild animals**
- **emotional factors, tradition and convention**

3.3 Gender in animals in the world's languages

This chapter will look into the gender of animal names in selected languages of the world, in which two major systems for assigning gender occur:

Semantic systems, "where semantic factors are sufficient on their own to account for assignment" (Corbett 1991: 8)

Formal systems, "where formal criteria determine gender to a large extent" (Corbett 1991: 37)

3.3.1 Basic types of assigning gender to animal nouns in languages with strict semantic or predominantly semantic systems

This part contains a summary of the issues connected with gender of animals in selected languages which, from the point of view of their gender assignment systems, belong to the same group as English does. It is a pronominal gender system, since gender is reflected only in personal, possessive and reflexive pronouns. The use of *he*, *she*, *it* pronouns is similar to those in Tamil, mentioned below. There is a high degree of variability with animals. Domestic animals, particularly if they are named, are masculine or feminine according to their sex. Moreover, many animals in children's stories have a particular gender by convention. (*cf* Corbett 1991: 12)

To sum up, it is the group with strict semantic or predominantly semantic systems of gender assignment. There are two basic types of information about the noun- its form and its meaning (semantics). These two basic types of information are necessary for the assignment.

The following summary is based on Corbett (1991).

Tamil and Dravidian languages

In **Tamil**, one of the major Dravidian languages, nouns are asssigned to three genders:

masculine (= male rational), feminine (= female rational), neuter (= non-rational) "animals may be treated as persons in fables, normally however, even where there are distinct words for the male and female of animals, all nouns referring to animals are neuter" (Corbett 1991: 9).

In **Kannada**, another language that belongs to the Dravidian family, the situation is similar to that in Tamil: "There is a very small number of exceptions to the semantic principles, words for bull and buffalo are masculine, gender reflects the special status of these higher animals" (ibid.: 10).

Zande belongs to the Niger- Congo branch. There are four genders: masculine, feminine, animal and neuter. "The first two genders are straightforward: nouns denoting male humans are of masculine gender and, equally, nouns of masculine gender denote male humans. Feminines are similar. The one minor complication is that for small children the pronoun for animals is used" (ibid.: 14)

In **Dyirbal**, an Australian language, "gender is marked on what Dixon calls 'noun markers', which are analogous to demonstratives" (ibid.: 106). Nouns can be assigned to four genders:

I. male humans, non-human animates
II. female humans, water, fire, fighting
III. non-flesh food
IV. residue

Animal nouns can be assigned to three genders, *e.g.*:

gender I – kangaroos, possums, bats, most snakes, most fishes, some birds, most insects
gender II – dog, platypus, echidna, some snakes, some fishes, most birds, firefly, scorpion, crickets
gender III – bees

There are a number of exceptions resulting from various associations (concept association, mythological associations, importance marking). The following quotation describes a clear example of an exception based on mythological associations:

"Thus birds (which are animate and so would be expected to be in gender I) are believed to be the spirits of dead human females: they are therefore in gender II But some individual birds have mythological associations which put them in gender I" (Corbet 1991: 16).

If nouns forming a subset in a particular category are distinguished by an important property, they may be assigned to a different gender. For example "fish" are in gender I. Two harmful species of fish, the stone fish and the gar fish, are set apart in gender II.(ibid.: 17). The reason is perhaps the association with fire and fighting.

In **Ket**, a language spoken in Siberia, all nouns denoting living things are of masculine or feminine gender but the division is sometimes hard to understand for non-sex-differentiable animates (where a single noun is used for both sexes). Corbett mentions possible explanations offered by Krejnovič that "those which show a higher degree of activity are masculine, but this hardly accounts for the fact that nouns denoting the following non-sex differentiables are feminine: hare, squirrel, chipmunk, rat, mouse and mole. Krejnovič believes that these are feminine because they are of no importance to the Kets" (ibid.: 19). Another interesting explanation concerning the two exceptions of fishes (fishes are referred to as masculine except for three fishes) "the ruff is not a particularly active fish, and the perch is not favoured as food by Kets" (ibid.: 20).

Lak and other **Caucasian languages** have four genders: male rational, female rational, other animate, and the residue. "The third gender comprises especially animates (*animals, birds, fish, insects*). The fourth gender includes very few animates (*spider, earwig, dragonfly*, also *butterfly* and *cat* in some dialects), some concrete objects, most liquids and almost all abstract nouns" (ibid.: 25).

Archi, another Caucasian language, has four genders. To gender III belong for example birds, domestic animals, larger wild animals, insects. Gender IV includes young animals and birds (wild and domestic), smaller wild animals and birds. "For animals sex is of no importance: the words for "cow" and "bull" are both in gender III There is a division between domestic animals and birds (all III) and wild animals and birds. The latter divide into larger (III) and smaller (IV), though exceptionally "mouse" is in III. The young of animals and birds are in IV...." (ibid. : 28).

In **Alamblak**, a language of Papua New Guinea, there are two genders – masculine and feminine. "Besides males, the masculine includes nouns whose referents are tall, or long and slender, or narrow such as fish, crocodile, long snakes,...The feminine comprises besides females, nouns denoting short, squat or wide entities: turtle, frog,..." (ibid.: 32).

The above overview includes the example of the influence of mythology on gender assignment (in Dyirbal). The influence of culture and mythology on gender assignment has been observed in many languages with semantic assignment systems.

English too has a gender system based on semantic criteria. It is a pronominal gender system. The use of *he*, *she* and *it* is determined by principles similar to Tamil, mentioned above. There is a high degree of variability with animals. "Domestic animals, particularly if they are named, are masculine or feminine according to sex; and, especially in children's stories, many animals have a particular gender by convention" (ibid.: 12). Above all, there are cases where the straightforward semantic rules are overridden by emotive and affective factors (Vachek 1964).

The above examples show how the meaning of a noun determines its gender in some languages while in others the role of semantics is more restricted.

In other languages, semantic criteria fail to account for the gender of a high proportion of the nouns, and formal criteria must be sought. Such formal criteria are the subject of the next chapter.

3.3.2 Assigning gender to animal nouns in languages with formal assignment systems

Even if the focus is predominantly on gender assignment in English, this chapter will provide a brief description of assigning gender in languages with formal systems to make the theoretical background more complete.

The following languages are under study:

German, French, Russian, Czech and Bantu.

In languages with a formal assignment system the assignment is not primarily based on semantic criteria but on formal assignment rules, that is, rules which depend on the form of the nouns. In formal systems, formal criteria (morphological in *e.g.* Russian, German, Bantu languages or phonological in *e.g.* French) determine the gender.

Most of the world's languages make use of mixed systems, but "gender always has a basis in semantics" (Corbett 1991: 63). This can be illustrated on the example of German:

German has a complex gender system with semantic, morphological and phonological assignment rules. If it is important, then some animals are distinguished by sex. In case of domesticated animals, there usually is a common noun (neuter) for both biological sexes and separate nouns for each sex and for a juvenile (neuter).

s Huhn (hen)	*r Hahn*	*e Henne*	*s Kuken*
s Pferd (horse)	*r Hengst*	*e Stute*	*s Fohlen*
s Schwein (pig)	*r Eber*	*e Sau*	*s Ferkel*

In most cases, one gender includes both biological sexes: *r Spatz (skylark), e Biene (bee).*

There are examples where the distinctions are made by a suffix, e.g. *r Leopard (leopard), -e Leopardin (leopardess).*

Phonological systems can be demonstrated on Afro-Asiatic, Kru and some Papuan languages. In Qafar, a language with two genders, the position of the accent is an indicator of gender. Numerous pairs like *kùta -(dog)* and *kutà -(bitch)* are given as examples in Corbett (ibid.: 51).

French is also a language where examples of the phonological aspects of assignment can be illustrated. In the majority of cases, there is only one form to refer to both members of the pair.

They might be of masculine gender, e.g.:

rat (rat), merle (thrush), serpent (snake), crapaud (toad), papilton (butterfly), moineau (sparrow), élephant (elephant),

or of feminine gender, e.g.:

grenouille (frog), araignée (spider), aloutee (skylark).

In cases when the nouns are distinguished morphologically, the masculine form has both generic and masculine specific meanings:

Tigre (tiger) – tigresse, chien (dog) – chienne, lion (lion) – lionne, loup (wolf) – louve, cheval (horse) – jument.

Russian, one of the East Slavonic languages, has three genders. Though it is said that Russian has a morphological assignment system, semantic factors still predominate. Agreement is shown by adjectives, verbs, pronouns (personal, possessive, relative). Semantic assignment rules are applied in sex-differentiable nouns, which means that sex- differentiable nouns denoting males -humans and some animals are masculine: *lev (lion).* Sex-differentiable nouns denoting females and some animals are feminine: *lvice (lioness).* Domesticated animals are often distinguished according to their sex, but in most cases the sex of animals is of no importance and so formal assignment determines the gender.

In Russian each of the three genders (masculine, feminine, neuter) is subdivided into two subgenders: animate and inanimate. Most neuters are inanimate but there are exceptions, e.g. nasekomoe (insects), and životnoe (animal).

Another group of animal nouns is formed by morphologically unrelated nouns, such as *byk* and *korova* (*bull and cow).*

The examples of standard Russian *kit (whale),* which denotes a male or female but takes masculine agreement in either case, and *akula (shark)* is, conversely, feminine, are used by Corbett (1991) to demonstrate the notion of epicene nouns and to distinguish them from common nouns. "Common nouns take two different sets of agreement forms, epicene nouns take only one, though they denote beings of either sex" (ibid.: 67).

"**Czech** represents a very typical specimen of the highly synthetic, inflexional kind of language" (Vachek 1964: 192).

It is the form, the derivation suffix and the type of declension that determine the gender assignment in Czech. Gender in Czech is predominantly a grammatical category and gender assignment is based on a formal system, thus an animal noun may belong to one of the three genders: masculine, feminine or neuter.

In animal nouns, gender assignment is not straightforward. Grammatical gender is conventional, based on the tradition and history of the language (Miko 1962: 27).

In Czech, when analysing nouns denoting animates and some animals, the natural semantic gender becomes the core component of the grammatical gender.

For non-native speakers of Czech it is not always easy to group animal nouns into masculine, feminine or neuter, especially in cases where the nouns have zero endings in the nominative singular case or endings untypical of the given gender.

Miko (1962) defines the role of natural gender as follows: "Besides the fact that natural gender forms the meaning core of grammatical gender, natural gender has its own grammatical traits that enable one to create a grammatical category" (ibid.: 37).

In Czech, the result of derivation is very often the change of gender: *medvěd – medvědice* (*male bear – female bear*).

In Czech the natural sex opposites, based on derivation, are limited in animal nouns. The female equivalents are used less frequently. They are used in well-known animals: *slon – slonice* (*male elephant – female elephant*), *vlk – vlčice* (*he-wolf – she-wolf*).

In the case of domesticated animals heteronymous pairs have been formed, such as *kachna – kačer (duck – drake), slepice – kohout (hen – cock), ovce – beran (ewe – ram), kočka-kocour* (*cat- tom-cat*).

The majority of animal nouns such as *pstruh (trout), zmije (viper), moucha (fly)* are unmarked from the point of natural gender, and they follow certain declension patterns based on whether they are masculine, feminine or neuter.

Gender differentiation and the asymmetrical use of animate nouns in contemporary Czech can be found in Dickins (2001: 212–247)

Bantu languages are said to have extensive gender systems. Assignment rules are based on both semantic and morphological principles and they are quite complex. The priority of certain semantic rules over morphological rules varies not only in languages but also in dialectal variations. Sex does not play any role in assignment. "In Swahili the rule applies to animates: in many other Bantu languages the rule is restricted to humans" (Corbett 1991: 48). For example the Swahili equivalent of "*giant snake*" belongs to gender 5/6, "*tiny snake*" to gender 7/8, "*lizard, rooster, rhinoceros, elephant, snake*" to gender 1/2. Other examples can also be found in Pinker (2000: 121–122).

4. Material Analysis

4.1 Introductory information

The aim of the analysis is to find answers to the following set of research questions:

- ***What are the criteria of assigning gender to animal nouns in English?***
- ***Are there any differences in assigning gender to animal nouns in the corpora explored?***
- ***How do Czech speakers assign gender to animal nouns in English?***
- ***To what extent do the differences between gender assignment systems in English and in Czech influence Czech speakers of English in their choice?***

Three sets of materials have been analysed.

- The first corpus represents 36 volumes of the English edition of *National Geographic* (1979–2015).
- The second corpus represents eight volumes of the Czech edition of *National Geographic* (2002–2009).
- The third corpus includes three volumes of the English edition of *Journal of Zoology* (2006–2008).

For my analyses I have chosen zoological articles concerning predominantly endangered species, taken from two different sets of materials: non-fiction and scientific literature.

- ***National Geographic Magazine***, later shortened to *National Geographic*, the official journal of the National Geographic Society, Washington, D.C. The first issue was published in 1888.
- As a life-member of *National Geographic Society* since 1979 I own the 36 years' collection of this magazine. Taking into account my degree in biology, the topic of this thesis seemed to provide an interesting field for exploration in gender studies
- ***Journal of Zoology***, a scientific journal concerning zoology, the study of animals. It was founded in 1830 by the Zoological Society of London and carries original research papers and reviews within the whole field of zoology.

To make the picture as complete as possible, a contrastive analysis of gender markers will be offered.

As an additional part of my research I have included the analysis of the use of gender markers by Czech students and native speakers.

To obtain the necessary data for my analysis I have used all gender markers in their respective contexts, retrieved from the corpora mentioned above. I have also used tests, translations from English to Czech and both quantitative and qualitative data analysis. The results of the research will also be presented in the forms of comments, charts and tables.

The research includes illustrative examples retrieved from the material. All the examples retrieved can be found in the appendices.

The research presented in this thesis is divided into the following chapters:

4.2. The analysis of gender of animals based on *National Geographic* articles
4.3. The analysis of gender markers in animals based on Journal of Zoology
4.4. The comparison of gender assignment in animals included in the sub-corpora of *National Geographic* and *Journal of Zoology*
4.5. The comparative analysis of gender in juvenile animals
4.6. The Comparative analysis of the use of gender markers by Czech students and native speakers

4.2 The analysis of gender of animals based on *National Geographic* articles

The corpus for this analysis includes all *National Geographic* articles concerning animal species published between 1979 and 2015.

I have retrieved all articles concerning animals ranging from insects to apes, concentrating on the identification of gender distinctions. Surprisingly, these occurred universally across all the species explored, including the lowest ones.

The following is the summary of animal species explored:

Insect
Species: Ant,
Earwig, Locust, Aphid, Fly, Ladybug, Flowerfly, Cricket, Dragonfly, Moth

Sea world
Marine Microfauna species
Marine fish species
Sea horse species
Shark species
Whale species

Reptiles and Amphibians
Snake species
Lizard species
Turtle species
Frog species

Birds
Birds of prey
Exotic birds

Big Cats
Lion species
Tiger species
Leopard species

Bears
Bear species
Panda species

Elephants

Apes

The objective of this analysis was to find out the rules and principles of gender assignment in animal nouns in the articles. I recorded all animal nouns with their gender references, which means all personal, possessive and reflexive pronouns referring to animal nouns in the corpus. Animals without gender references were not listed.

The animals are divided into classes and species, given in brackets. Each animal group and its referring pronouns are recorded in a separate table. Each table is followed by material analysis, including sentences quoted in italics as examples. The other examples, referring to the respective animal species can be found in the appendices. To make the gender marking clearer, I divided all the pronouns retrieved according to their gender and highlighted them in the text. The retrieved examples are further divided according to their reference in the articles (e.g. examples referring to reproduction, parental care, life cycle, behaviour, observation or personal interest of the researchers).

The results of the analysis of each animal group are given in a separate chart.

I have divided animals into classes according to their biological classification.

The following subchapters present the analyses of all the animal classes mentioned above, starting with the "lowest" class **Insects** and its species:

4.2.1 Insect

To this class belong the following species retrieved from the corpus: **Ant, Earwig, Locust, Aphid, Fly, Ladybug, Flowerfly, Cricket Weta, Dragonfly, Moth, Mosquito, Beetle and Praying Mantis.**

The insect species are divided into two parts. The first part covers the ant species, which belong to the most frequently mentioned insects in the corpus. The second part includes the rest of insect species retrieved from the corpus.

4.2.1.1 Ant

(Camponotus Ant, Bulldog ant, Marauder ant, European ant, Australian sugar ant, African stink ant, Nothomyrmecia ant, Leafcuttler ant, Pseudomyrmex ant, Crematogaster ant)

Table 4.1: Gender assignment – Ants

Pronouns	Masculine		Feminine		Neuter	
Personal subjective	he	3	she	35	it	4
Personal objective	him		her	15	it	
Possessive	his	2	her	60	its	9
Reflexive	himself		herself	4	itself	
TOTAL		5		114		13

<u>Material analysis</u>

10 ant species were referred to by a total of 132 pronominal pronouns in the following percentages:

Masculine: 5 = 3.8%
Feminine: 114 = 86.4%
Neuter: 13 = 9.8%

Total: 132 = 100%

Figure 4.1: Distribution of gender – Ants.

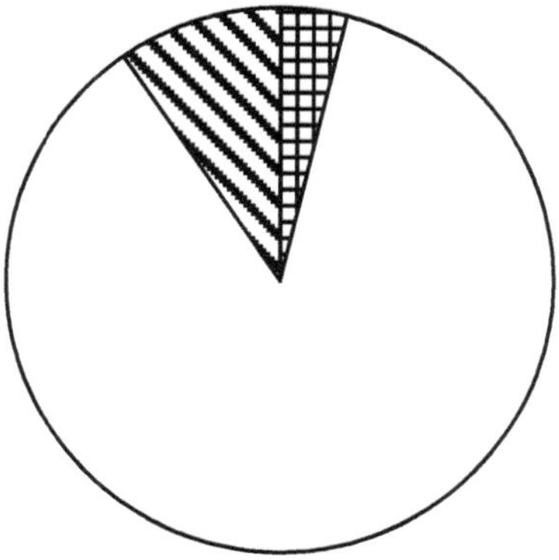

A quantitative analysis has revealed that the dominant gender was feminine – 86.4% of all pronouns, only 3.8 % masculine and 9.8 % neuter.

The dominance of feminine gender may be explained by the social life of this animal.

Ant colonies are female societies, where all workers are female. Consequently the ant world is a mother-and-daughter universe (*National Geographic* July 1995). In general, the ant itself is mostly viewed as **she** (cf. A.4.–A.25.) Even the headline of one of the main articles concerning ants "***The Ant** and **Her** World*" (*National Geographic* June 1984, p.775) supports this hypothesis.

The most important member of the ant colony is the queen. The queen is always referred to by feminine pronouns. The majority of quoted sentences describe the queen's life in the colony, her reproduction, egg laying and influence over the workers. (cf. A.2.–8, A.9.–12.,A.15.–17.). The following examples describe the life cycle and reproduction process of the queen:

(1) [A.6.] When **she** and the local guard rubbed antennae, both would snap their abdomens skyward, open their jaws, and begin jerking back and forth like sparring boxers, releasing alarm pheromones. **Her** greatest labour ahead, **the mated harvester queen** lands and breaks off **her** now useless wings. Working alone, **she** is vulnerable to predatory birds, lizards, and other ants; only one in 100 to 1,000 young queens survives these first days.
Quickly the queen digs **her** nest with mandibles and forelegs, making balls of the moist soil to lift around the entrance. Once enchambered below the surface, **she**

lays eggs. When they hatch into larvae, **she** feeds **her** brood on secretions from metabolized fat reserves and wing muscles.
When the colony's population reaches several thousand, the queen produces winged males and females. Truly **she** is queen of all **she** surveys.
(*National Geographic* June 1984, p. 793)

(2) [A. 8.] Only **the queen**, who may boast a thousand times the body weight of a minor worker, is fertile. All workers are **her** daughters. Occasionally **she** spawns new queens, who fly off to mate and start new colonies. **Her** wings shed, this hopeful marauder monarch attaches **her** newly laid eggs to **her** body and protects them by hobbling along with **her** back section scrunched forward.
(*National Geographic* August 1986, p. 275)

The ant colony makes a suitable setting for gender distinctions because quite a number of expressions describe gender roles in it. All members mentioned are characterized as ***she*** in all the examples retrieved: *a sister* (cf. A. 20., 22.), *a mother* (cf. A. 14.), *a huntress* (cf. A. 21.), *a hunter* (cf. A. 19.), *a soldier* (cf. A. 24.), *a worker* (cf. A. 11., 13., 18.)

I have chosen the following examples to show the gender assignment of a worker, a soldier and a mother:

(3) [A. 18.] With a leaf in **her** jaws, **a worker** cannot defend **herself**, so small ants ride shotgun to keep flies at bay.
(*National Geographic* July 1995, p. 110)

(4) [A. 24.] I watched **one soldier** poise **herself** to cut a bloody arc into my skin as effortlessly as a smaller forager would cut leaves.
(*National Geographic* July 1995, p. 108)

(5) [A. 14.] Rushing toward the silky white nest, **mother ant** extends a gland from **her** depressed abdomen. Like a pack of dogs, the weavers surround the giant and subdue **her** with formic acid.
(*National Geographic* June 1984, p. 783)

The feminine gender is also used in the case of a close connection between the animal and the author. I have found, for example, that emotional involvement of the researcher becomes particularly apparent in situations when the ants have been observed over extended periods of time.

(6) [A. 21.] My pride was a colony I raised when I was 14 from a single queen I had caught in the fields right after **her** nuptial flight. **She** was still alive when I left home at 25. I must admit I was somewhat disappointed with the behaviour of Nothomyrmecia. As Bob Taylor observed, this ancient ant is fully social. **She** is not the hoped-for missing link between social and solitary behaviour.
(*National Geographic* June 1984, p. 785)

Generally, other genders in the ant colony are marginal.

I have found only five examples of masculine gender. They describe the life cycle of the drone (cf. A.3.) and his role in the mating process (cf. A.2.). Unlike the female, the male has only one role – mating with the queen. Shortly thereafter he dies (*National Geographic* June 1984), as described in the following example:

(7) [A. 1.] The **drone**'s role ends with the frenzied glory of the nuptial flight. **He** dies shortly thereafter, burned out and literally consumed from within by **his** brief, energy intensive life of sperm production.

(*National Geographic* June 1984, p. 793)

That is why the occurrence of masculine pronouns in the corpus is so low.

The usage of neuter pronouns is also surprisingly low and typically connected with the life cycle of the ant (cf. A.25.–28.) and its observation. One possible explanation of neuter markers usage in the following example is the general description of life cycle of the particular ant species:

(8) [A. 29.] I never saw **a marauder ant** searching for food on **its** own. **The Asian jumping Ant**, known for **its** ability to make great starling leaps, may travel as far as a hundred feet from **its** nest- a long journey for an ant. But finding prey makes it worthwhile. This jumping ant in Hong Kong has seized an unwary cockroach with **its** long, curved mandibles.

(*National Geographic* August 1986, p. 278)

Another possible explanation of neuter gender markers usage is the individual approach of the authors.

4.2.1.2 Earwig, Locust, Aphid, Fly, Ladybug, Flowerfly, Cricket Weta, Dragonfly, Moth, Mosquito, Beetle, Praying Mantis,

Table 4.2: Gender assignment – Insects

Pronouns	**Masculine**		**Feminine**		**Neuter**	
Personal subjective	he	2	she	28	it	5
Personal objective	him	2	her	12	it	
Possessive	his	13	her	52	its	19
Reflexive	himself		herself	1	itself	1
TOTAL		17		93		25

Material Analysis

12 species of insect were referred to by a total of 135 pronominal pronouns as follows:

Masculine:	17 = 12.6%
Feminine:	93 = 68.9%
Neuter:	25 = 18.5%
Total:	135= 100%

Figure 4.2: Distribution of gender – Insects.

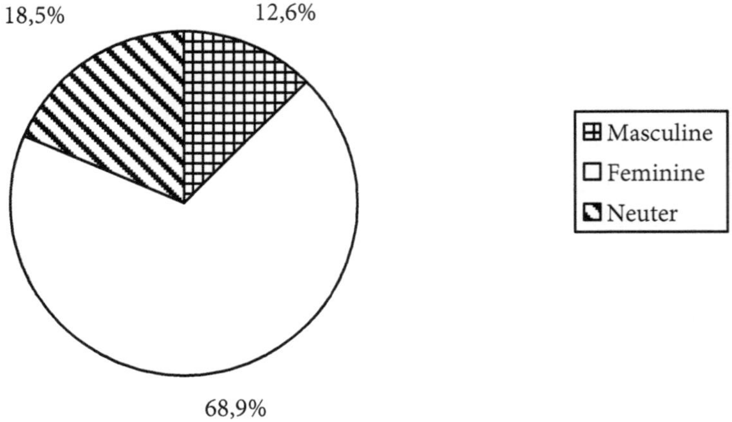

It is surprising, that such a "low" animal as an aphid distinguishes gender roles.

The dominance of feminine pronouns in the animals mentioned above is closely connected with:

- parental care, particularly the relationship between mother and her young. The personal interest of the authors is manifested in using comparisons and expressions labelled in the examples mentioned below. Authors use expressions such as *a young, a newborn, offspring, a hatchling, a baby and grandchildren* to express youngsters. This can be documented by the following examples describing maternal care in earwig and aphid:

(9) [A.45.] After eggs hatch, **the female earwig** watches over **<u>her</u> young** for as long as four weeks. **The hatchlings use their mother as chicks do a brooding hen**, scapering beneath **<u>her</u>** body at the slightest sign of danger.

(*National Geographic* August 1980, p. 284)

(10) [A. 46.] Emerging from **<u>her</u> mother**, a Pseudoregma soldier reaches down with **her** legs to pull **<u>herself</u>** free. Moments later **the mother seems to cradle <u>her</u> offspring**, but in fact **<u>she</u>** provides no parental care. **<u>Her</u>** next birth may be either a normal aphid or another soldier.. The non-soldier embryos inside **<u>her</u>** body already contain **<u>her</u>** developing **grandchildren**. This is one reason why aphids develop so rapidly.

(*National Geographic* September 1989, p. 412)

- the process of mating. When mating is described, explicit gender distinctions are used in all cases. I have excerpted gender distinctions referring to reproduction in the moth (cf. A.33.,35.), the fly (cf. A.36.,37.), the ladybug (cf. A.38.,39.) and the dragonfly (cf. A.40.–44.).

The most illustrative example is the mating process of the **praying mantis**. As mating proceeds, the female clutches the male around the neck and begins to feed on his head. Though now decapitated, the male has a nervous system that enables him to continue mating. Some scientists have suggested that the male thus benefits his offspring by providing protein for the female during egg production. (*National Geographic* 1984, vol. 165, p. 274, cf. A.29.–32.,34.)

(11) [A. 31.] The female twists **<u>her</u>** head and begins to eat **<u>his</u>** head and forebody during the act. Having fulfilled **<u>his</u>** primary role, the male now is most useful to **<u>his</u>** species as food.

(*National Geographic* 1984, vol. 165, p. 276)

The neuter gender assignment is marginal. Only 25 neuter pronouns were found, referring to the description of life cycle (cf. A.67.–76.) and three neuter pronouns referring to youngsters (cf. A.65., 66.). Although authors use the neuter forms when describing the life cycle and behaviour of these individuals, their personal interest is apparent. They use comparisons as documented by the example adduced below:

(12) [A. 67.] **As still as the stone, a rock locust** also matches **<u>its</u>** granite perch with camouflaging color.

(*National Geographic* August 1980, p. 566)

<u>Insect- Final Chart</u>

Masculine:	22 = 8.3%
Feminine:	207 = 78.1%
Neuter:	38 = 13.6%
Total:	267 = 100%

Figure 4.3: Final chart – Insect.

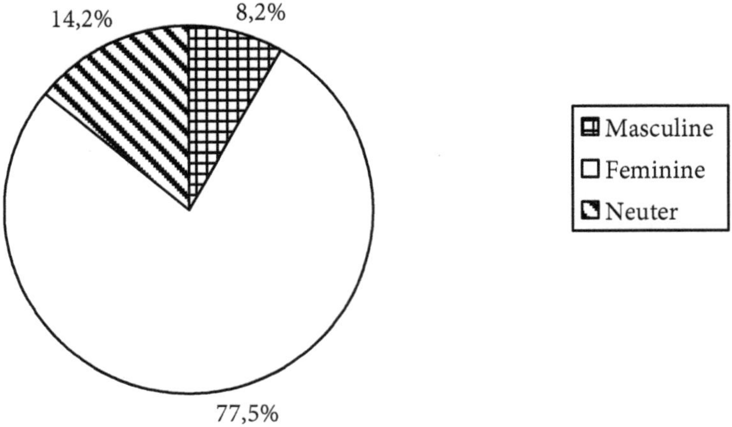

Overall counts suggest that **in the world of the insect the feminine gender is dominant**. The explanation of this fact is simple. The female plays the main role in the reproductive process, egg laying and maternal care. That is the reason why authors pay attention predominantly to females, in 78.1 % of all examples.

4.2.2 Sea Species

The following part of my research presents the analysis of marine species. I have divided them into three groups:

- **Marine microfauna**
- **Marine fish species**
- **Shark and whale species**

4.2.2.1 Marine microfauna

(Amphipod, Shrimp, Jellyfish, Squid, Octopus, Cuttlefish)

Table 4.3: Gender assignment – Marine microfauna

Pronouns	Masculine		Feminine		Neuter	
Personal subjective	he	3	she	3	it	28
Personal objective	him	1	her		it	5
Possessive	his	11	her	14	its	66
Reflexive	himself	2	herself		itself	3
TOTAL		17		17		102

Material analysis:

Masculine: 17 = 12.5%
Feminine: 17 = 12.5%
Neuter: 102 = 75.0%

Total: 132 = 100%

Figure 4.4: Distribution of gender – Marine microfauna.

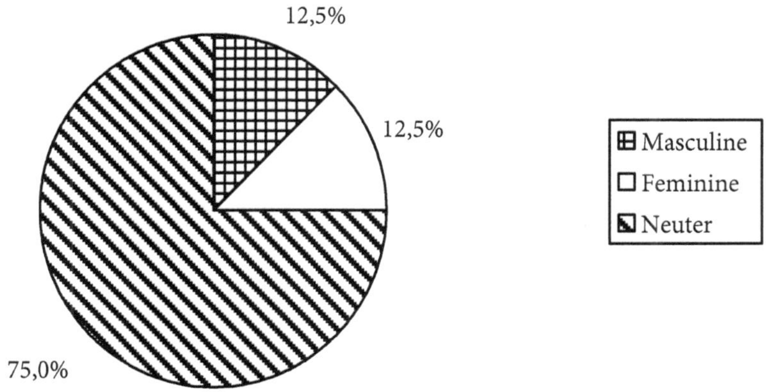

As can be seen from the examples above, the authors distinguish the masculine or feminine gender in 25% of the excerpted samples of the marine species. The distribution of masculine and feminine gender is balanced- 12.5% of all pronouns were masculine, 12.5% feminine. The balance may result from parental roles of these animals.

Surprisingly, the authors distinguish the gender even in "low" animals, such as squid, cuttlefish or jellyfish. Even though these species belong to low biological classes of *Cephalopoda, Scyphozoa and Malacostraca*, the gender distinction is not parallel with the biological classification. The authors thus view them as higher animals.

This fact is evident under the following circumstances:

- in the description of the **mating process** in the squid (cf. B.10.), the octopus (cf. B.17) and the cuttlefish (cf. B.27.) as described in the following examples:

(13) [B.10.] A red-armed male embraces a female after mating. **She** will lay more than 50,000 eggs.

(*National Geographic* August 2004, p. 40)

(14) [B.17.] Mating can take hours. The male has enveloped the female's mantle with **his** web.
During mating the heclocotylus is inserted into the female's mantle cavity, and a spermatophore is attached near one of **her** oviducts. In **her** den a female stares out from beneath **her** nest, which contains as many as 80,000 eggs. Using a mucous secretion, **she** attaches them to ceiling in strands of 150 to 200 eggs.
Always on guard, the female aerates the eggs by shooting water through **her** siphon and removes bits of foreign matter with **her** arms. For six months or longer **she** perseveres, rarely if ever eating, and nearly always dies of starvation. The crabs and small fish were feeding on **her** eggs.

(*National Geographic* March 1991, p. 94)

(15) [B.27.] When squarting off with another male, a large adult ripples **his** gangly outer arms, thereby emphasizing **his** size. Another male, cruising limestone overhangs for a female to mate with, elongates **himself** to look intimidating. Attracted by the displays of the suitor, **she** proceeded to mate, even as one of the spurned males spread **his** arms and bared **his** beak in an empty threat. The mating proceeded quickly. The male used one of **his** arms to deposit sperm capsules into a pouch beneath the female's mouth. The capsules burst open, spreading sperm into **her** mantle and fertilizing the eggs. The coupling completed, the female swam back into the cave to lay **her** new eggs.
What shows clearly is **her** siphon. By expelling water through the funnel-like tube, a fleeing cuttlefish can propel **itself** backward in a surprising burst of speed.

(*National Geographic* September 1995, p. 105)

- in situations, where **parental care** is described. The authors use the expressions, such as *mother, baby, newborn* when describing the relationship between a mother and her young. I have retrieved two species with feminine and masculine gender assignment- in amphipod (cf. B.1.,2.) and shrimp (cf. B.4.,5.). Even if these crustaceans range from one millimetre in length and thus belong to lowest individuals in marine microfauna (*National Geographic* December 1981), they are referred to by ***he*** and ***she,*** as can be seen in the following examples:

(16) [B.1.] **A female amphipod** devoured a salp's internal organs and saved the casing to serve as in incubator and **baby carriage** for **her** soon-to-be-hatched larvae.
(*National Geographic* December 1981, p. 842)

(17) [B.5.] As with most crustacean, because of their rigid exoskeletons the male must wait to implant **his** sperm until the female has molted **her** old shell.
A screen prevents **the mother** from eating **her** own eggs, which appear as small bubbles in **her** feeding basket. Colleagues of the author discovered that a female can produce many sets of eggs during the summer season.
(*National Geographic* May 1984, p. 636)

The gender distinction is particularly evident in cuttlefish. This animal has a highly developed brain for invertebrates; consequently *he* or *she* is very often viewed as a "higher" animal. The personal involvement of the authors is evident (cf. B.26.,28.):

(18) [B.26.] Off Yorke Peninsula, near Adelaide, I came upon an individual that had matched **himself** to the sandy floor of **his** hideout. When **he** saw me, **he** turned bright yellow. Did my strobe excite **him**? Or did **he** change colors simply to see what I would do?
(*National Geographic* September 1995, p. 98)

In 75% of all examples the neuter form is used. It is obvious in general descriptions of life cycle (cf. B.3.,6.–9.,11.–16.,18.–22.,30.–33.), observation (cf. B.23.–25.) and youngsters (cf. 29.)

I have found one example where the shrimp observed is named "Charlie" (cf. B.7.). Despite of the fact that the animal has its own name, the authors surprisingly have not referred to it as ***he***:

(19) [B.7.] Only one, whom we named **Charlie,** would eat the prawns we placed in **its** mouth. The other jellies died, while **Charlie** doubled in size within a month.
(*National Geographic* August 1994, p. 118)

4.2.2.2 Fishes

(Stargazer, Batfish, Jawfish, Goby, Flag benny, Coffinfish, Pearlfish, Flounder, Sea moth, Scorpionfish, Parrotfish, Waspfish, Mola, Cichlid, Anemonefish, Clownfish, Surgeonfish, Fridmanfish, Humphead, Anglerfish, Triggerfish, Hawkfish, *Foetorepus altivelis*, Cutlass, Bristlemouth fish, Lizard fish, Dragonet, Lionfish, Sandager's wrasse, Stargazer, Harlequin tusk fish, Angelfish, Anthias, Pouchy-mouthed fish, Bathylagid, Mola mola, Chinook, *Tyrannochromis macrostoma,* Damsel, Wrasse, Devilfish, Barracuda, Sculpin, Triplefin)

Table 4.4: Gender assignment – Fishes

Pronouns	Masculine		Feminine		Neuter	
Personal subjective	he	8	she	6	it	36
Personal objective	him	2	her		it	2
Possessive	his	11	her	13	its	70
Reflexive	himself		herself		itself	5
TOTAL		21		19		113

Material analysis:

Masculine: 22 = 13.7%
Feminine: 19 = 12,4%
Neuter: 113 = 73.9%

Total: 153 = 100%

Figure 4.5: Distribution of gender – Fishes.

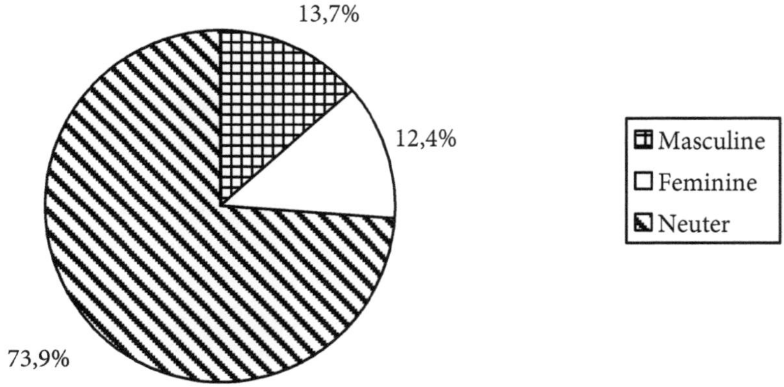

In the case of **fish** species the use of **neuter** gender is dominant- 75% of all pronouns were neuter. While the majority of animal species play only one role (masculine or feminine), some are described in two or three roles.

I have excerpted 44 tropical fish species. Many of them have a very special social structure, mating process and parental care. This is the main reason why authors distinguish the gender roles when referring to the specific fish species. For example the individuals of the *Anthias* and *Anemone* fish species are able to transform themselves into a male or a female during the reproductive process (cf. B.40.–43.):

(20) [B.40.] They are called anthiases. The female's the little orange one, the male's purple one. They're the ultimate survivors. One male services a whole crowd of females. If the male dies or disappears, if **he**'s eaten by something, the dominant female will begin a gradual transformation into a male. **She**'ll become a male and will be able to breed with females.

(*National Geographic* April 1988, p. 449)

(21) [B.41.] Deep maroon marks a spinecheek anemonefish as the dominant female of this host anemone, where **she** rests with **her** orange mate. When **she** dies, **he** will turn maroon, become female, and mate with the next male in line. This color-coded sex change vividly shows that in the drama of coral reefs, color plays a leading role.

(*National Geographic* May 2005, p. 108)

The authors also use feminine and masculine gender in the descriptions of parental care, in which both "parents" play important roles. The example below describes the parental care of *Flag benny* species, where "the father blenny" – as the authors call him – plays a very important role in the protection of his young (cf. B. 34.,36.,38.,39.):

(22) [B.36.] I sit back and wait for **him** to settle down in **his** new home. Thirty minutes pass before **he** takes on the job of wriggling **his** body over the eggs to aerate and clean them. While **the father blenny** protects the eggs, I reset the camera for a new series of shots.

(*National Geographic* July 1992, p. 107)

Even more interesting is the following description of maternal care:

(23) [B.39.] Bearing the burdens of motherhood, a female Tyrannochromis macrostoma guards **her** brood as they search for food. At the first sign of danger **she** chases away an intruder. **She** then signals **her** young to reenter **her** mouth, where they are out of harm's way. Releasing **her** eggs one at a time, the female takes the egg into **her** mouth, then nuzzles the male's genital region, causing **him** to release sperm into the water, which the female inhales, thereby fertilizing the egg. The female carries the eggs in **her** mouth until they develop into juveniles, protecting them until they are ready to survive alone.

(*National Geographic* May 1990, p. 47.)

To sum up, the roles of both male and female marine fish in parental and mating process are remarkably balanced and similarly balanced is the distribution of masculine and feminine gender marking – 12.2% and 12.7% respectively.

Instead of the description of reproductive processes and parental care, the authors distinguish the gender roles when describing the life cycle (cf. B. 35.,42.,43.) and observation (cf. B. 36.,37.,44.) of marine fishes. In the example below the author describes the finding of the rare, deep water fish. Consequently he marks the observed individual as ***he***:

(24) [B.37.] **It**'s a bathylagid. This is probably the first time anyone's seen this species alive in **its** own habitat. Look at how big **his** eyes are. **He** feeds on all the gooey stuff in the water. It may be that **he** looks up to see it. The manipulator arm accidently touches the bathylagid, and **it** darts away.

(*National Geographic* February 1990, p. 39)

The occurrence of neuter pronouns in marine fishes is high. 73.9 % of the retrieved gender markers are neuter. They are connected with the description of their life cycles (cf. B. 50.–84.), observation (cf. B. 85.–94.) and youngsters (cf. B. 47.–49.)

4.2.2.3 Sea horse

(H. abdominalis, H. ingens, H. hippocampus, H. comes, H. histrix, H. zosterae, H. erectus, H. reidi, H. funcus, H. breviceps, H. whitei, H. bargibanti)

Table 4.5: Gender assignment – Sea horse

Pronouns	**Masculine**		**Feminine**		**Neuter**	
Personal subjective	he	7	she	3	it	
Personal objective	him	6	her	7	it	
Possessive	his	18	her	1	its	
Reflexive	himself	1	herself		itself	
TOTAL		32		11		0

Material analysis

When referring to 12 species of seahorse, the authors used pronominal pronouns as follows:

Masculine: 32 = 74,4%
Feminine: 11 = 25,6%
Neuter: 0 = 0%

Total: 43 = 100%

Figure 4.6: Distribution of gender – Sea horse.

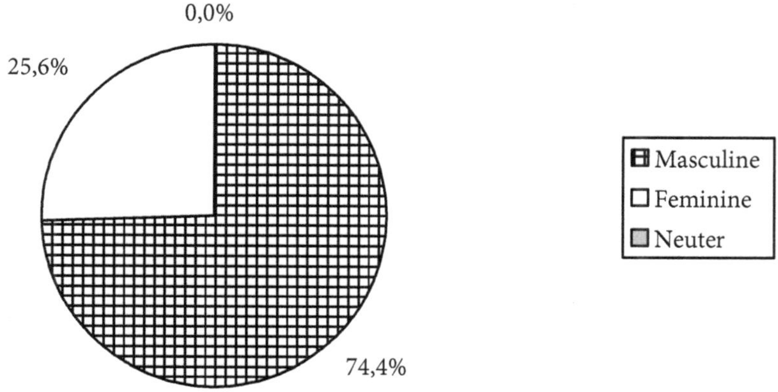

Even though the sea horse belongs to the group of fish, I have paid special attention to this animal. There are over 32 species of sea horses. From the *National Geographic* articles I have adduced 12 species. As can be seen from the examples and the chart depicted above, the occurrence of pronouns is very unusual- the authors have not used neuter pronouns. Thus the sea horse is the only animal with the dominance of masculine pronouns.

It could be explained by the specific behaviour of this animal. Males play a very important role in the reproductive process -they get pregnant, give birth and do all the parenting (cf. B.95.,98.,99.–101.). In the example below the author compares the seahorse's pregnancy to mammals' and even the youngsters are marked as *babies*:

(25) [B.99.] In addition to reinforcing the pair bond, greetings probably cue a female to **her** mate's impending delivery, enabling **her** prompt preparation of a new clutch of eggs. From start to finish a male seahorse does all the parenting. **His is indeed a real pregnancy, closely resembling that of mammals, including humans.** Just as a woman provides her fetus with oxygen, transfers nutrients, and regulates conditions in the uterus, a seahorse does no less for the embryos in **his** pouch. **His** pouch was only half a tablespoon in volume, but in it were packed 1,572 **babies**; end to end they would have stretched more than 12 yards.

A newly single female stays in **her** home range, whereas a bereft male starts roaming once **his** pouch is empty. We can be sure of seahorse fidelity because a female's body visibly deflates when **she** transfers eggs, while the male's pouch inflates.

(*National Geographic* October 1994, p. 135)

The authors also distinguish the gender assignment in the observation of these individuals (cf. B.96.,97.,102.). In both examples the male and the female have numbers:

(26) [B.97.] My first target is often **male number 35. His** ballooning pouch makes it awkward for **him** to weave among the sea grasses, and **he** may cling to the same shoot for days. Yet so well camouflaged is **he** that I sometimes spiral around **him** several times before spotting **his** tail, the only part visible from all angles. Brushing a vivid orange, a male H. reidi- recently delivered of young- twirls with **his** partner in a courtship ritual that may last nine hours, when **he** again becomes pregnant.

(*National Geographic* p.139, October 1994, p. 139)

(27) [B.102.] We are both waiting for **his** mate to appear, but few things hurry a seahorse. At last **female 10** meanders toward **him**. The male's home range is about a square yard, whereas **his** female's can be a hundred times as large, thus reducing competition between them for food. **She** swims past other males en route, but they ignore each other. Each male is awaiting **his** own female.
When **female 10** nears **her** pregnant partner, sparks fly- or at least colors change.

(*National Geographic* October 1994, p. 134)

For all the reasons mentioned above, 74.4% of retrieved pronouns are masculine and only 25.6% are feminine. There is no need to use the neuter gender.

4.2.2.4 Shark

(Tiger shark, Great hammerhead, Lemon shark, Great white shark, Greenland shark, Horn shark, Nurse shark)

Table 4.6: Gender assignment – Shark

Pronouns	Masculine		Feminine		Neuter	
Personal subjective	he	4	she	10	it	20
Personal objective	him		her	11	it	1
Possessive	his	10	her	19	its	39
Reflexive	himself	1	herself	1	itself	5
TOTAL		15		41		65

Material analysis

When referring to 7 shark species the authors used pronominal pronouns as follows:

Masculine:	15	=12.4%
Feminine:	41	=33.9%
Neuter:	65	=53.7%
Total:	121	=100%

Figure 4.7: Distribution of gender – Shark.

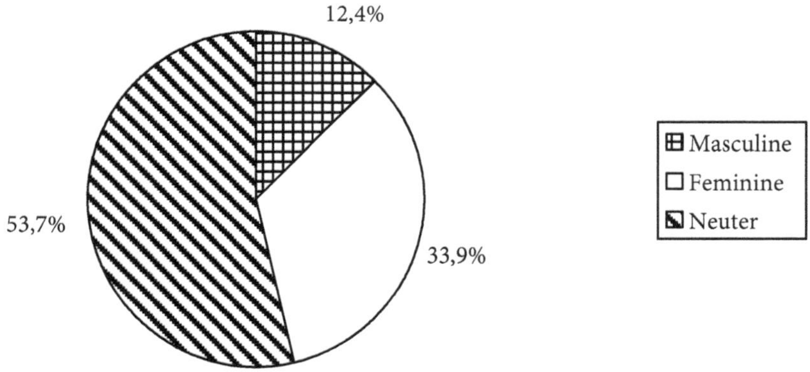

Although studies of behaviour and the reproductive process of this type of fish offer little information, I have nevertheless observed examples where sharks have been documented and observed in detail.

The occurrence of feminine gender is evident- 35% of all pronouns were feminine.

The **masculine and feminine gender** was predominantly used when the authors wanted to describe the specific roles of the shark:

When the courtship and the mating process are described, the explicit gender distinction is used. (cf. B.106.–B.110). Although mating has rarely been observed in sharks (*National Geographic* May 1995), I have retrieved an example describing the sex life of sharks in the wild. In documenting such a unique situation, scientists marked the female and her partner as ***he*** and ***she***:

(28) [B.107] Locked in a turbulent embrace, two nurse sharks mate off the Florida Keys- the first time scientists have documented in detail the sex life of sharks in the wild. To subdue **his** partner, the male must seize the female's pectoral fin, flip **her**, and carry **her** from the shallows to deeper water.

(*National Geographic* May 1995, p. 45)

The next example describes the courting rituals:

28. (B.109.) Violent **courtship** stirs the Red Sea. Bitten by a male, a female short-nosed gray reef shark shows gashes from dorsal to caudal fins. **Her** companion, a large jack, may be riding **her** slipstream.
Almost coyly, a female peels off alone. A male rushes **her** and bites **her** fins and flanks. After the attack the two then usually return to the group.
After biting a female, **her** suitor speeded up, wriggled over **her** back, then swam briefly along **her** right side.

(*National Geographic* August 1981, p. 159)

The gender distinction is used in observation. One of the sharks studied has his own name- *Red Eye*. This dominant male has been referred to by masculine pronouns in all excerpted cases:

(30) [B.104] **His** stare as fierce as **his** bite, the dominant male we called **Red Eye** prepares to haul away the female, remoras and all. **Red eye** probably scratched **his** eye on the rocky bottom while wrestling for position.

(*National Geographic* May 1995, p. 50)

If the shark is referred to emotively and with personal interest, then it is viewed as ***he*** or ***she***. All excerpted examples referring to observation, fishing and diving showed the tendency to personalize the shark and treated the individual as ***he*** or ***she*** (cf. B. 105., 110.–114.). In the examples below researchers who have made many dives with sharks all over the world describe their experiences with this fish:

(31) [B.105.] But that day I was spearfishing in a big competition. With lots of fish blood in the water, it must have been too much for the shark. I'm just lucky **he** only bit me and let go. If **he**'d bitten down and shaken me in the feeding technique, I wouldn't be here." Although I have made hundreds of dives with sharks around the world, I was particularly excited at the thought of this one.

(*National Geographic* August 1981, p. 183)

(32) [B.111.] The big female that breaks formation and heads my way passes so close I can make out the pores that pepper **her** snout and enable **her** to sense the electromagnetic energy of living flesh. As **she** slides by, huge and silent, I reach out and run a hand over **her** side. **Her** movements stay steady and calm and **she** rejoins the circling sharks.

(*National Geographic* March 2007, p. 128)

From the data cited above the preference of neuter gender is evident. The authors mainly use **neuter gender markers** in descriptions of shark observation (cf. B. 122.–129.)

(33) [B. 129.] <u>**It**</u> turned and began moving toward me. I was face-to-face with a Greenland shark. <u>**It**</u> was ghoulish. <u>**Its**</u> nostrils were the largest I had ever seen on a shark. <u>**Its**</u> mouth was slightly open. <u>**Its**</u> eyes looked fogged over.... The shark showed renewed interest in me. Afraid <u>**it**</u> would suck me into <u>**its**</u> menacing maw, I backpedaled furiously toward the surface.
I still wanted to document fully this eerie predator in <u>**its**</u> underwater lair.
(*National Geographic* September 1998, p. 62)

The shark is also marked as ***it*** in the examples referring to life cycle (cf. B. 116.–121.).

Juveniles are referred to by ***it*** in 100% of retrieved examples. The authors used the expressions as *a nurse shark, a baby, a newborn, a pup* when referring to youngsters (cf. B. 113., B. 116., B. 119., B. 120.).

4.2.2.5 Whale

(Beluga whale, Minke whale, Humpback whale, Bottlenose whale, Sperm whale, Harbor porpoise)

Killer Whale ("Orca")

Table 4.7: Gender assignment – Whale

Pronouns	**Masculine**		**Feminine**		**Neuter**	
Personal subjective	he	10	she	18	it	16
Personal objective	him	8	her	6	it	1
Possessive	his	18	her	49	its	32
Reflexive	himself		herself		itself	2
TOTAL		36		73		51

<u>**Material analysis**</u>
When referring to 7 whale and orca species the authors used pronominal pronouns as follows:

Masculine: 36 = 22.5%
Feminine: 73 = 45.6%
Neuter: 51 = 31.9%

Total: 160= 100%

Figure 4.8: Distribution of gender – Whale.

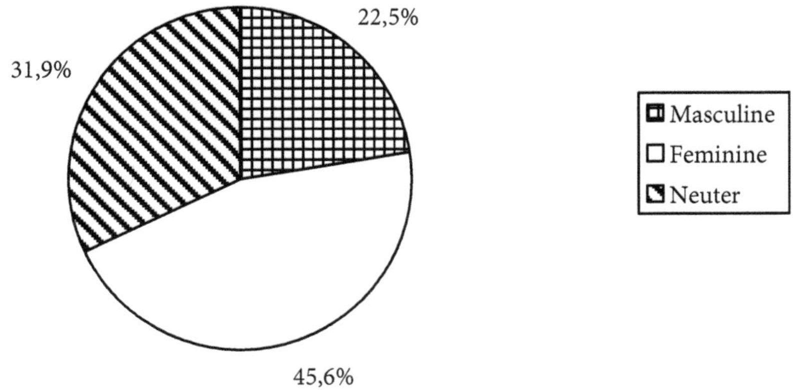

These marine mammals belong to the group of the most frequently described animals in *National Geographic*. I have divided these marine animals into two groups- **whale** and **orca** species. Orcas are referred to as killer whales, but according to their biological classification they belong to dolphins. I have found a number of articles concerning whales and orcas. One half of the excerpted samples distinguish the gender and its roles. The preference for masculine and feminine gender in the articles is evident – 64.7% of all examples. It can be explained by the fact that these animals belong among seriously endangered species, thus the authors personalise them and pay special attention to them.

The dominance of the feminine gender can be seen in most of the examples depicted below.

This fact is connected with the behaviour and life cycle of these mammals- females are often spotted in groups, called pods: "*They consist of as many as four generations of related whales of both sexes- **great-grandmothers, grandmothers, mothers, sisters, and their offspring, including mature males".*** (*National Geographic* December 1988, p.892)

The most frequently-mentioned relationship in the articles is between the mother whale and her calf. All the above-mentioned facts might help to explain that 43.3% of all recorded pronouns were feminine. The examples concern the family structure. The authors personalise the members of the pod and use terms similar to humans. Consequently all the retrieved female individuals are referred to by feminine markers (cf. B. 137.,138.,140.,142.,143.,144.,151.,152.,):

(34) [B.156.] "Most **cows** with **calves** are accompanied by another adult, which scientists had long assumed to be **a female "aunt"** that helped the other raise **her calf.**"

(*National Geographic* December 1988, p. 901)

(35) [B.142.] A calf will squirm over and under **its** mother's flippers, or pectoral fins-**her** modified forelimbs. When **she** lingers at the surface, **it** will use **her** massive back as a playground. An infant that wriggles onto **her** rostrum, the broad bench formed by the elongated upper jaw, seems happy to perch there like some colossal amphibian equally at home in water on in air, while the mother half shoves, half carries **her** bulky baby. However, the majority of them have at least one male suitor following or, from time to time, angling ahead or closely behind as if to alter **her** course. When a female rests underwater, **her** visitor stations himself nearby, breaking off to patrol around **her** now and then, on the lookout for potential rivals. One afternoon, 50 feet deep in dazzling azure, a baby humpback rests tucked beneath **its** mother's flipper, then moves to nestle in **its** next favourite place: under **her** throat.

(*National Geographic* January 2007, p. 87)

The whales are very often given names such as *Sharky*, *Grace* and *Scar*. Each name has been chosen according to the appearance or behaviour of each individual. Two examples illustrate the author's attitude to the animal. The first male was named after the author's son. The female, described in the second example obtained her name because of her friendly behaviour. All named animals are therefore viewed as ***he*** or ***she*** (cf. B. 135., 136., 137., 141.):

(36) [B.135.] One of these they named **Owen**, after my youngest son. I watched with pride as **he** darted off into the green water, and later I followed **his** progress on the Internet until **his** transmitter gave out. I imagine **him** making deep dives in the dark Atlantic, chasing silver walls of herring. I wonder what perils **he**'ll encounter and if **he**'ll enjoy a long and fruitful life.

(*National Geographic* June 2003, p. 77)

(37) [B.137.] Mary Lou's whale was promptly dubbed **Amazing Grace**, for **her** friendly behaviour at the time was somewhat amazing and **her** movements underwater, like those of all **her** species, were graceful beyond description. From our very first encounter with **Amazing Grace**, **she** readily adopted us along with our 14-foot inflatable outboard as **her** personal toys. **She** would roll under the boat, turn belly up with **her** flippers sticking three to four; feet out of the water on either side of the craft, then lift us clear off the surface of the lagoon, perched high and dry on **her** chest between **her** massive flippers. When **she** tired of the benchpress technique, **Grace** would do the same thing with **her** head, lifting us out of the water and letting us slide off to swirl around **her** in circles, like a big rubber duck in the bathtub with a ten-ton playmate. At other times **Grace** would submerge beneath us and release a tremendous blast of air that boiled to the surface in a giant Jacuzzi of white water that engulfed us and the boat. After such gymnastics **Grace** would

often lie quietly alongside the boat to be rubbed. We would oblige **her** with a vigorous massage along **her** back, head, and ribs, while **she** opened **her** mouth to display huge fringed curtains of creamy white baleen plates.
From **her** length of 30 feet and the mammary slits along the after part of **her** belly, we judged **Grace** to be a two-year-old juvenile female. **She** would begin to mate and calve at puberty, about six years later.

(*National Geographic* June 1987, p. 755, 758)

On the other hand, the individuals included in the scientific research are usually marked for easy identification. All the retrieved animals are assigned to feminine. The absence of neuter in the case of research studies is surprising. Generally scientists have referred to the observed animal to by ***it***.
(cf. B. 140., 141., 149., 150., 154., 158.):

(38) [B. 149] **A resident female named J8** is known by **her** stubby fin. **She** heads one of several family trees, or matrilines, within the J pod. The septuagenarian **J8** and **her** offspring have entertained generations of whale-watchers.

(*National Geographic* April 2005, p. 93)

When describing courtship and mating strategies, I have found predominantly masculine gender markers (cf. B. 131., 133., 138.):

(39) [B. 131.] One male has bested about 15 rivals in a boisterous skirmish to court the female below **him**. "I have also seen two males pancake the third," Salden says. "One came up under **him** while the other lunged onto **his** back"

(*National Geographic* January 2007, p. 85)

The occurrence of neuter gender markers is connected with the description of whales' life cycle (cf. B. 159.–168.), observation (cf. B. 173.–179.) and youngsters (cf. B. 169.–172.).

Juveniles are also viewed as ***he*** and ***she***. I have found four young with gender distinctions (cf. B. 128., B. 145., B. 155., 165.). The gender in whale and orca juveniles is analysed in detail in the chapter referring to youngsters.

Sea Species – Final chart

All marine species mentioned above were referred to by a total of 568 pronominal pronouns as follows:

Masculine: 84 = 14.8%
Feminine: 160 = 28.2%
Neuter: 324 = 57%

Total: 568 = 100%

Figure 4.9: Final chart – Sea species.

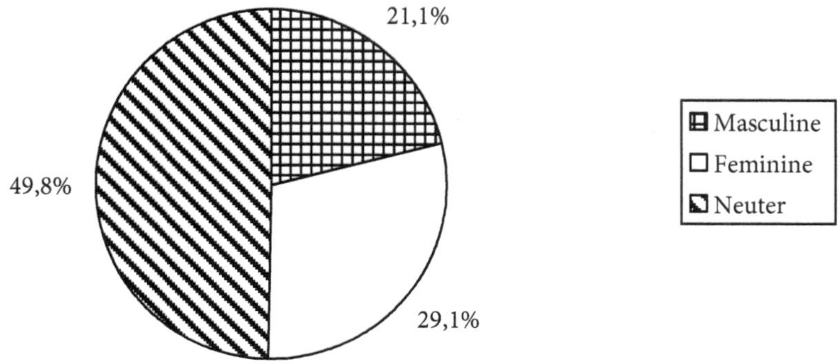

The above chart shows the balanced occurrence of definite and indefinite pronouns when referring to all retrieved marine species.

The animals with the lowest occurrence of definite pronouns were those belonging to the **marine microfauna**. The **jellyfish** was the only animal referred only to by neuter.

The prevalence of neuter gender markers is clear in **fishes** (73.9% pronouns). Only one third of the marine fishes retrieved have been marked with *he* or *she*.

More than half of neuter markers have been found in **shark** species.

On the other hand, **whale** and **orca** species belong to marine animals with the prevalence of feminine and masculine pronouns.

The only marine animal with the absence of the neuter form was the **sea horse**.

To sum up, marine species are viewed predominantly as *he* or *she*.

4.2.3 Reptiles and Amphibians

4.2.3.1 Snake

(Adder, Eastern diamondback Rattlesnake, Anaconda, King cobra)

Table 4.8: Gender assignment – Snake

Pronouns	Masculine		Feminine		Neuter	
Personal subjective	he	1	she	5	it	10
Personal objective	him		her	4	it	
Possessive	his	1	her	6	its	7
Reflexive	himself		herself		itself	
TOTAL		2		15		17

4.2.3.2 Lizard

(Angolosaurus lizard, Draco lizard, Green Basilisk lizard, Short-Horned lizard, Water- skimming lizard, Gecko, Chameleon, Tuatara, Agama, Brandt's lizard, Iguana, Mulgara, Thorny Devil)

Table 4.9: Gender assignment – Lizard

Pronouns	Masculine		Feminine		Neuter	
Personal subjective	he	1	she	2	it	8
Personal objective	him		her		it	
Possessive	his		her	5	its	23
Reflexive	himself		herself		itself	1
TOTAL		1		7		32

4.2.3.2 Turtle

(Alligator, Snapping Turtle, Green Sea Turtle, Hawksbill Sea Turtle, Leatherbask Sea Turtle, Loggerhead Sea Turtle, Olive Ridley Sea Turtle, Coahuilan box turtle)

Table 4.10: Gender assignment – Turtle

Pronouns	Masculine		Feminine		Neuter	
Personal subjective	he	3	she	16	it	9
Personal objective	him	4	her	7	it	3
Possessive	his	6	her	16	its	9
Reflexive	himself		herself	4	itself	
TOTAL		13		43		21

4.2.3.4 Frog

(D.auratus, D.pumilio, Brachycephalus sp., Hylodes asper, Scinax arduous, Agalychnis callidryas, Poison-arrow frog, Horned frog, Túngara frog, Harlequin Frog)

Table 4.11: Gender assignment – Frog

Pronouns	Masculine		Feminine		Neuter	
Personal subjective	he	6	she	19	it	4
Personal objective	him	2	her	1	it	2
Possessive	his	7	her	24	its	8
Reflexive	himself	1	herself	1	itself	
TOTAL		16		45		14

Material analysis:

The lizard, snake, turtle and frog species were referred to by a total of 226 pronominal pronouns as follows:

Masculine: 32 = 14.2%
Feminine: 110 = 48.7%
Neuter: 84 = 37.2%

Total: 226 = 100%

Figure 4.10: Distribution of gender – Reptiles and Amphibians.

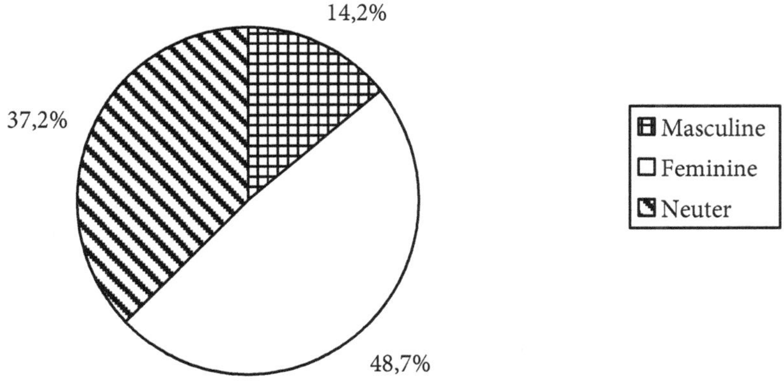

In this part I have put together two biological classes – **Reptiles** and **Amphibians.**

Reptiles include the lizard, snake and turtle species, amphibians covers frog species.

I have found 21 species of reptiles and 10 species of amphibians.

The number of articles concerning snake and lizard species was lower in comparison with the articles about turtles and frogs. Turtle and frog species belong to frequently mentioned animals in the corpus.

The above data show that the preference of feminine gender in these two classes is high – 48.7% of all excerpted pronouns are feminine.

In **snake and lizard species** the occurrence of feminine markers is connected with the observation (cf. C. 2.–4., 13.). The next example shows the life story of the female snake observed. The researchers marked her with number and studied this individual for years:

(40) [C. 4.] **Number 38** had weighed 82 grams and measured 52 centimetres long in 1978 when **her** two-segment rattle showed **her** to be a yearling. **She** measured 92 centimeters in length and might be ready to produce young. I gently passed my thumb along **her** abdomen but felt no lumps. **She** lacked developing eggs. This year, at least, **she** wouldn't bear young. The saga of number **38** reached a climax in 1986, when I captured **her** for the third time and found **her** pregnant in **her** tenth year…
(*National Geographic* July 1987, p. 131)

The occurrence of **masculine** markers in **snakes** and **lizards** is very low. Although I have found four snake and thirteen lizard species with gender distinctions, I have discovered only three masculine markers, referring to observation of trapping a snake male (cf. C. 1.) and a lizard male (cf. C. 12.):

(41) [C. 12.] Texas horned lizard appears poised for takeoff on the runway. Actually **he**'s intent on catching insects, mainly ants.
(*National Geographic* February 1981, p. 172)

On the other hand, the preference for the **feminine in turtles and frogs** is closely connected with the main role of females in the reproductive process and maternal care. Even if these species belong to different classes according to biological classification, the gender assignment principles are very similar. It is the female which is most important in the reproductive process and egg laying both in turtles and frogs. Therefore it is the female which is studied in the majority of examples and referred to as ***she***. (cf. C. 32.–34.) Detailed description of egg laying in turtles provides the following example:

(42) [C. 34.] Black and gleaming in the moonlight, each female drags **herself** from the surf, front flippers scoring the sand as **she** pulls **herself** along, and settles in to dig. Scooping with **her** rear flippers, **she** excavates a shaft. When **she** can no longer reach the bottom, **she** begins to lay **her** eggs, a glistening cue ball every few seconds. Once **she** has a cache of 80 or so, **she** fills in the nest, sweeping **her** front flippers to smooth out the spot. Then **she** drags **herself** a few feet away and makes

more giant sand angels-a decoy nest that may serve to confuse predators. After two or three hours on the beach, **her** throat rosy with exertion, **she** returns to the sea.

(*National Geographic* May 2009, p. 130)

The occurrence of masculine pronouns in turtles is connected with the reproduction (cf. C.29.–30.)

A similar gender assignment can be seen in **frog species**. The female plays the main role in reproductive process and eggs laying (cf. C.43.–50.):

(43) [C.45.] A dominant male pushes an immobile rival away from the female at rear. After fertilization, the female is left to guard **her** batch of 2 to 16 eggs hidden in leaf litter on the forest floor. When the tadpoles hatch, **she** backs in among them, and one wriggles up onto **her** back. On a rare trip away from the forest floor **she** climbs the canopy. **She** seeks little pools of water cupped in the leaves of certain plants. Into one of these pools **she** deposits the first tadpole. Then **she** goes back for another. One by one **she** carries **her** tadpoles to private water holes in the canopy.

(*National Geographic* May 1995, p. 106)

The following example describes the main role of the female in maternal care:

(44) [C.49.] The female poison dart frog not only attends **her** eggs but also transports the hatched tadpoles to pool in tree holes. **She** visits **her** offspring to lay unfertilized eggs for them to eat. The female Surinam toad converts **her** entire back to a nursery. **She** and the male swim end over end in their mating dance, transferring the fertilized eggs to pouches in **her** back. **Her** skin grows over the eggs, sealing them in until the froglets hatch.

(*National Geographic* May 2001, p. 113)

In frogs a great diversity of parental behaviours have been identified. Amphibians create the most interesting range of parental care I have discovered. This could be the reason for the highest feminine and masculine gender occurrence within these two biological classes. The animals are referred to with a clear personal interest of the researchers, who use the terms *father*, *dad*, *mother* when describing a variety of behaviours. Even for the frog youngsters the authors use a variety of expressions such as *tadpole, toadlet, froglet* and *young*. In the following example the male takes care of his froglets. The authors mark him "*Dad*" and the male is referred to by masculine in all depicted cases (cf. 36.–42.):

(45) [C.36.] That's a male's nest, and those little clear balls are the eggs **he**'s guarding- or should be, if **he** was there." **Dad** has apparently stepped out, Felton thinks. Their **froglets** climb aboard **dad**'s back and ride with **him** for a few days.

(*National Geographic* May 2001, p. 112)

As can be seen from the above charts, the authors, surprisingly, have not used **neuter** as the dominant gender in reptiles and amphibians. The neuter has been mainly used for general descriptions of life cycle, diet and observation. The highest occurrence of neuter markers has been found in lizards (cf. C. 14.–28.) while the lowest were among frogs (cf. C. 51.–60.). The use of the neuter form in the general description of the diet in one of the frog species shows the example below:

(46) [C. 59.] As a tadpole this frog survived dragonflies, water beatles, fish, and shrimps. As an adult <u>**it**</u> must evade bats, birds and snakes.

(*National Geographic* May 2001, p. 113)

4.2.4 Birds

(Sea Eagle, Harpy Eagle, Philippine Eagle, Nothern Goshawk, Snowy Owl, Great Grey Owl, Whooper swan, White Pelican, Albatross, Raven, Heron, Quetzal, Bee-Eater, Bird of Paradise, Cock-of-the-Rock, Hornbill, Jacana, Trumpeter, Penguin, Prairie Boomer, Turkey)

Table 4.12: Gender assignment – Birds

Pronouns	**Masculine**		**Feminine**		**Neuter**	
Personal subjective	he	27+47	she	23+9	it	5+20
Personal objective	him	9+11	her	6+5	it	1
Possessive	his	38+73	her	23+23	its	9+43
Reflexive	himself	2+2	herself	4	itself	1+2
TOTAL		209		93		81

<u>**Material analysis:**</u>
In the class of birds I have found 20 species with gender distinctions.
I have divided this class into two main groups:
Birds of prey
Exotic birds

4.2.4.1 *Birds of prey*

Birds of prey- especially eagles- are referred to by a total of 143 pronouns as follows:

Figure 4.11: Distribution of gender – Birds of prey.

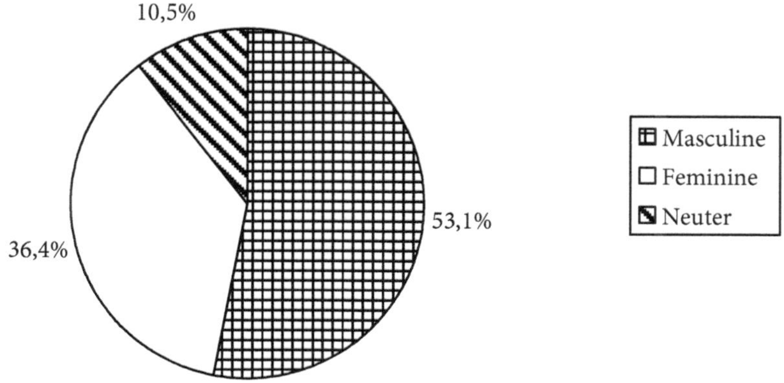

I have excerpted **Eagle, Goshawk** and **Owl** species, which belong to birds of prey. All articles concerning birds of prey describe the life of these birds, their training for special purposes such as hunting and the personal attitudes, experience and exploration of the researchers.

The **eagle** was the most frequent species of birds in the articles. I have found three eagle genera (Sea, Harpy and Philippine Eagle), all of them with gender distinctions. The dominance of articles concerning this bird may result from the fact that eagles, together with other birds of prey, belong among endangered species. Moreover, the eagle plays an important cultural role as the national symbol of many nations. Consequently, the majority of examples refer to observation of these endangered individuals. Both males and females have been observed. All individuals explored are referred to by ***he*** or ***she*** (cf. D. 11.–15., D. 16.–25.) as in the following example describing the observation of the female eagle and her young:

(47) [D. 11.] I did not know, where the powerful **mother eagle** was. "Will **she** attack?" I kept asking myself, and half hoped **she** would. For I was the intruder, and **her eaglet**, on the face of it, surely must have seemed in critical danger. "Here **she** comes!" At last moment **the eagle** pulled out of **her** attack." Here **she** comes again!" With **her** great talons, the irate **mother eagle** raked my shoulder. The scar on the helmet was proof enough of the eagle's maternal instinct to defend **her** nest.

(*National Geographic* June 1981, p. 847)

As can be seen from the above example, the researchers observe these birds very closely. Every bird has its own name or mark. Even youngsters are marked and viewed as ***he*** or ***she***. In general, the adult eagle is more often viewed as ***he***. As can be seen from the chart below, the dominant gender in birds of prey is masculine- 53.1%. *It is especially the male, who provides food and protects his family.* (*National Geographic* February 2005, cf. D. 6.–10.):

(48) [D.9.] Meanwhile, the male keeps bringing plucked birds to **his** mate.
(*National Geographic* July 1994, p. 125)

Both the male and the female take part in parental care. I have found examples, where the male plays the main role in parental care and fledging (cf. D. 1.–5.).

The preference of feminine gender is evident in the descriptions of feeding the youngsters, egg hatching and fledging (D. 16.–27.). The authors describe in detail the feeding of an eagle female and her young. The personal involvement of the researchers is clear:

(49) [D.16.] "12: 41- The female began feeding **herself**. **She** ate 37 pieces of meat, offering several to the eaglet. At 12: 50 the chick began to regurgitate after considerable struggle-suddenly **it** began thrusting in death throes; the baby eagle died a minute later."
(*National Geographic* June 1981, p. 850)

I have also found interesting examples referring to eagle training for hunting purposes. All the trained birds were females, referred to by ***she*** (cf. D. 32.–34.):

(50) [D.32.] A veteran hunter strains to support his eagle as **she** flattens **her** wings, several feet from tip to tip. He often hunts from this promontory and when his eagle pursues prey in the valley below, he jumps on his horse and races down the mountain, hoping to reach **her** before **she** damages the precious fur or the animal injures **her**.
(*National Geographic* March 1999, p. 93)

The neuter in birds of prey is marginal, for it covers only 10.5% of retrieved pronouns. They are used in the context of parental care and nesting (cf. D. 35.–40.), the life cycle (cf. D. 41.–43.) and observation (cf. D. 44.). In the case of such specific and endangered species, there is no need to use neuter gender markers. Generally, the authors prefer more personal and emotive ***he*** and ***she*** when referring to these birds.

4.2.4.2 Exotic birds

1. **Bee-Eater, Bird of Paradise, Cock-of-the-Rock, Hornbill, Jacana, Kakapo, Kiwi, Puffin, Trumpeter, Quetzal**
2. **Albatross , Heron, Prairie Boomer, Penguin Raven, Turkey Whooper swan, White Pelican**

The quantitative analysis has revealed that the dominant gender in the second group of birds is masculine – 56.7% of all pronouns. The number of neuter pronouns is higher in "exotic birds" in comparison with birds of prey.

Figure 4.12: Distribution of gender – Exotic birds.

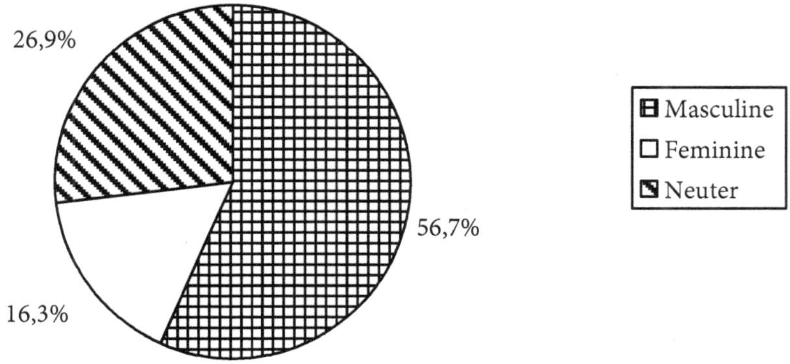

I have named this group of birds "exotic" to distinguish them from birds of prey. In fact, only the first group mentioned above belongs to the group of so – called exotic birds. The low occurrence of the neuter could be connected with their exotic origin. The majority of these birds are very rare and threatened with extinction. Consequently, they require finer gender distinction. Examples could be given of New Zealand's "exotic birds" such as kiwis, kakapos and snared crested penguins, which belong to this specific group. The authors have not used neuter markers when referring to these species.

The 100% occurrence of masculine and feminine marks in the articles could be explained by the fact that all kakapos have names, many of them suggested by New Zealand children. I have analyzed the articles of one of these New Zealand's ancient parrots of the night named *Aranga*. Her name means "resurrection" in Maori. (*National Geographic* p.83, October 2002) The authors described her as a pet:

(51) [D.82.] I was helping weigh a three-year-old **female kakapo** caller **Aranga**. (All kakapos have names, many of them suggested by New Zealand children. Aranga means "resurrection".) **She** used **her** beak as an ice ax and **her** toes as crampons. As **she** nibbled the sweet potato I was using to coax **her** onto the scale, I felt the tickle of **her** whiskers and the warmth of **her** down. When the food was gone and **Aranga** looked up with **her** inquisitive owl-like face, I had to laugh.

(*National Geographic* October 2002, p. 83)

I have found an example referring to the behaviour of a male kakapo who is also referred to entirely by masculine pronouns:

(52) [D.63.] There was a **kakapo**, New Zealand's ancient parrot of the night. **He** was standing in **his** bowl, a shallow depression **he** had made in the peaty soil. **His** calls carry up to three miles.

(*National Geographic* October 2002, p. 80, 81)

To sum up, the **masculine** in this group of birds dominates. Masculine gender distinction is evident in examples

- Referring to life cycle and behaviour (cf. D. 55.–64.)
- Referring to observation (cf. D. 65.–73.)

Both masculine and feminine is used in descriptions

- Referring to parental care, nesting and incubation (cf. D. 45.–54.)

Feminine pronouns are used in examples

- Referring to parental care (cf. D. 74.), life cycle (cf. D. 75.–80.) and observation (cf. D. 81.–85.)

The following example is a beautiful model of behaviour and parental care in jacana, observed in a National Park. The authors personalise the birds and mark them as ***he*** and ***she***, respectively:

(53) [D.52.] A male pheasant-tailed jacana strutting on lotus leaves near **his** floating nest in Yala East National Park. The female deserts **her** mate and seeks a new partner. For 26 days the male incubates the eggs, folding **his** wings like a cradle beneath them. Then **he** shepherds **his** new brood for about three months. **The father** eyes **his** newly emerged **firstborn**.

(*National Geographic* August 1983, p. 264)

Another example of personal involvement is the description of the raven:

(55) [D.68.] "For me **Raven** is a crest in my family lineage, **he** is an ancestor; **he** is a transformer who can turn **himself** into a hemlock needle; **he** is a trickster; or **he** is a human. We still relate to **his** power through our dances and songs."

(*National Geographic* January 1999, p. 109)

To sum up, gender distinction in this group of birds is very common. Instead of gender markers, authors use expressions which are specific for birds. For example, in trumpeters they distinguish ***pen*** for a male and ***cob*** for a female (cf. D. 80., D. 81.).

To describe their family life, the authors used the terms as *mother, father, chick, baby*. Even youngsters were referred to by ***he*** and ***she***.

The use of neuter is marginal. It covers only 26.9 % of all pronouns. It refers to parental care (cf. D. 86.–94.), life cycle (cf. D. 95.–111.) and observation (cf. D. 112.–113.).

Birds – Final chart

All above mentioned bird species were referred to by a total of 383 pronominal pronouns as follows:

Masculine:	209 = 54.6%
Feminine:	93 = 24.3%
Neuter:	81 = 21.1%
Total:	383 = 100 %

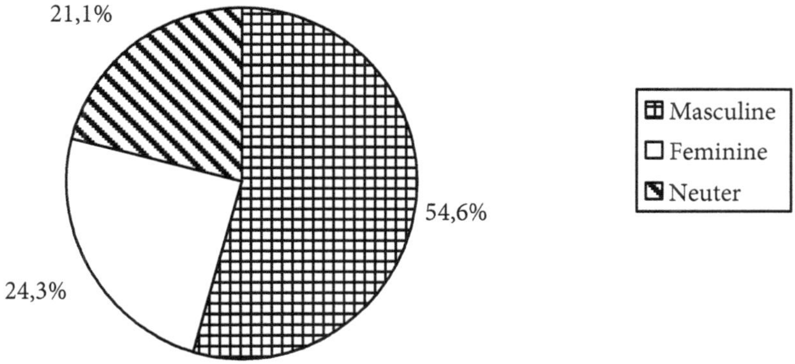

To sum up, **the dominant gender in assigning birds is masculine.** It dominates especially in the descriptions of a wide range of birds' behaviours and life cycle.

4.2.5 Big cats

The next group of animals, which has been frequently described in the *National Geographic* corpus, belongs to the family of "Big Cats". This group includes the four main species of cat in the genus of *panthera*- the **lion, tiger, leopard** and **jaguar**, which are quite similar both in behaviour and structure (*National Geographic* October 2001). I have paid attention to gender distinctions in the most frequently- mentioned big cats in the corpus- lion, tiger and leopard and their subspecies given in brackets. I found no gender references concerning the jaguar.

4.2.5.1 Lion

(East African Lion, Mountain Lion, Indian Lion)

Table 4.13: *Gender assignment – Lion*

Pronouns	Masculine		Feminine		Neuter	
Personal subjective	he	11	she	16	it	
Personal objective	him	5	her	4	it	2
Possessive	his	12	her	23	its	
Reflexive	himself		herself		itself	
TOTAL		28		43		2

<u>**Material analysis:**</u>
The three lion subspecies mentioned above are referred to by a total of 75 pronominal pronouns as follows:

Masculine: 28 = 38.4%
Feminine: 43 = 58.9%
Neuter: 2 = 2.7%

Total: 75 = 100%

Figure 4.13: Distribution of gender – Lion.

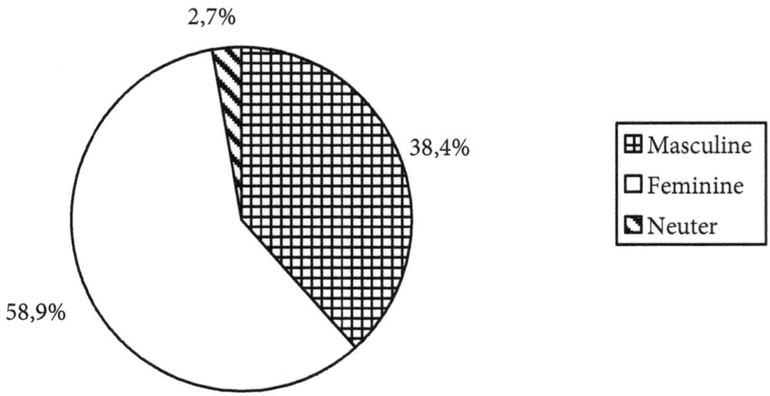

The lion belongs to the species very often described in the articles of *National Geographic*. As can be seen from the above chart, feminine gender marks dominate. This dominance is based on the hierarchy of lions. It is the lioness which plays an important role in "the lions' society". Related females and offspring create the pride of lions. Very close relationship can be seen between the lioness and her cub (cf. E. 10.–14.):

(55) [E. 11.] "Ever so carefully, a lioness slides **her** enormous canine teeth around **her** five-week-old cub's head before picking up the youngster. **She** carries **it** 15 feet, then puts **it** down and moves on ahead, calling all the while with soft, moaning grunts. Finally the cub responds to **her** coaxing and trots after **her**.
(*National Geographic* December 1982, p. 811)

The mother lioness very often has her own name (cf. E. 13., 14.):

(56) [E. 13.] There **Catrina** bore **her** kittens. **Catrina** leaps on a year- old kitten- now nearly **her** own size- **her** jaws landing squarely on its back, where mountain lions often seize their prey. In another lion lesson **she** pins a kitten to the ground.
(*National Geographic* July 1992, p. 48)

On the other hand, not only females take care about their cubs. In the following example a male lion is described as a very tolerant parent: (cf. E. 3., 4.):

(57) [E. 3.] "Cub play begins around dawn with the little ones climbing all over the sleeping male, pummelling **him** with their paws, chewing **his** tail, and tugging at **his** mane so hard that they extract bits of hair. Finally **he** rouses and walks over to a sleeping lioness while a cub clings valiantly to **his** tail."
(*National Geographic* December 1982, p. 817)

The female lion provides food for her pack. Female lions usually hunt together in groups, but I found only descriptions of a single female's hunting skills and strategies, as mentioned in many examples. The lioness has been always viewed as ***she*** (cf. E. 15., 17.–20.):

(58) [E. 15.] "Accelerating, **she** closes to within a few yards of the wildebeest. At this point we though **she** was assured of a kill, but three successive pictures record a slip, a fall, and a miss. Even in the act of falling, **she** kept **her** eyes riveted on **her** quarry. After returning-with an obvious limp to **her** ambush site, the lioness waited while more zebras gathered to drink. **She** charged, but missed again. Afterward **her** cub licked **her** face as if in commiseration with **her** failure."
(*National Geographic* December 1982, p. 807)

Many examples were found describing unsuccessful hunting strategies (cf. E. 16.):

(59) [E. 16.] "This lioness did not get kicked, but the hooves made **her** falter. **She** missed the kill and was left standing in a cloud of dust. Another lioness was not so fortunate.

Kicked while the pride chased and killed a zebra, the left side of **her** jaw flops askew, broken. Standing over the kill, **she** is unable to eat. We never saw **her** again.
On an earlier occasion we photographed another lioness with a broken jaw, presumably from a zebra's kick. One of the park rangers found **her** dead at a water hole ten days later."

(*National Geographic* December 1982, p. 819)

Male lions do not participate in the hunt. All retrieved examples concern only the diet (cf. E. 8., 9.):

(60) [E. 8.] "According to folklore, only an old or injured lion will eat a porcupine, with the ingested quills **his** final meal. But it was evident that this lion had eaten them before. **He** knew exactly what **he** was doing. When we saw **him** months later, **he** looked fresher and stronger, supporting a recent study that found that healthy lions in the Kalahari Desert include porcupines in their diet."

(*National Geographic* December 1982, p. 815)

In the descriptions of the reproduction process, both male and female lions are referred to by ***he*** and ***she***. The researchers give lions names (cf. 1., 2.):

(61) [E. 1.] "In the mating game, the lioness calls the signals. The lioness circles the male-giving **him her** scent-flicks **her** tail in **his** face, and walks away-growling. **He** follows and they mate for several seconds. Then there's a brief spat. They snarl and **she** may cuff **him**. After ten or fifteen minutes **she** initiates mating again."

(*National Geographic* December 1982, p. 808)

(62) [E.2.] We called **him** **Burr Boy** because of the burrs mating **his** scruffy side- whiskers. **He** was courting **his** female dining companion, **Melinda**, as the second lioness…

(*National Geographic* April 2002, p. 39)

I found only two examples of neuter markers, both of them connected with the youngster (cf. E. 11.).

4.2.5.2 Tiger

(Siberia's tiger, Wild tiger, Sumatran tiger, Bengal tiger)

Table 4.14: Gender assignment – Tiger

Pronouns	**Masculine**		**Feminine**		**Neuter**	
Personal subjective	he	55	she	50	it	
Personal objective	him	9	her	11	it	
Possessive	his	36	her	53	its	
Reflexive	himself	2	herself	3	itself	
TOTAL		102		117		0

Material analysis:

The four tiger subspecies mentioned above were referred to by a total of 222 pronominal pronouns as follows:

Masculine:	102 = 46.8%
Feminine:	117 = 53.2%
Neuter:	0 = 0%
Total:	222 = 100%

Figure 4.14: Distribution of gender – Tiger.

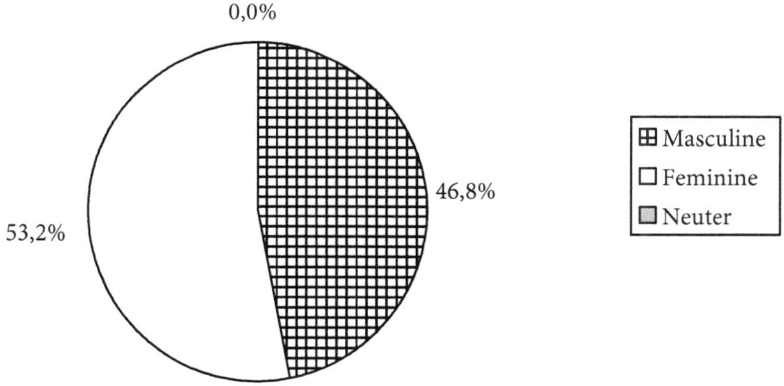

As can be seen from the examples adduced below, the gender assignment in tigers is closely connected with their life cycle. They are solitary, territorial animals. That is why the majority of examples contain one particular individual. Nearly all the tigers observed have names (cf. E. 21.–69.):

(63) [E. 64.] "We had known the cubs' mother, **Lena**- **<u>she</u>** was one of our study animals. (*National Geographic* July 1993, p. 39)

As the researchers say, "*we name the tigers we study*" (*National Geographic* p. 751, December 1984).

The above chart shows the balance of masculine and feminine gender markers:

In the descriptions of parental care, it is the mother tiger which is mentioned (cf. E. 21., 22., 46.–62.):

(64) [E. 62.] "Perhaps **<u>she</u>** recognizes me from our earlier encounter, for **<u>she</u>** hisses, spits and snarls with **<u>her</u>** lips curled back. Banseri slumps, drapes **<u>herself</u>** over the rock and sleeps. In late January we notice as **<u>she</u>** lies beside **<u>her</u>** kill that **<u>she</u>** has swollen

teats. **<u>She</u>** has cubs! A month later the family moves down the slope to a small, flat area of bamboo. Banseri lies on **<u>her</u>** side. First one cub, then the other, comes over to rub faces with **<u>her</u>** and to nurse.

(*National Geographic* December 1984, p. 769)

Both mothers and youngsters have very often names and are thus referred to as ***he*** or ***she***:

(65) [E. 54.] "**Kulu** and **<u>her</u>** cubs wade in a water hole in Kanha. As **a mother**, a tigress must by **<u>herself</u>** find food and train **<u>her</u>** offspring for about two years until they become independent.

(*National Geographic* December 1984 p. 757)

In the descriptions of mating, tigers are always referred to by ***he*** and ***she***. I have found two examples concerning the mating process. The animals have names (cf. 41., 42.):

(66) [E. 41.] "**Kulu** no longer lashes **<u>her</u>** tail. **<u>She</u>** lies down in the sand. Then **<u>she</u>** gets up, yawns widely, stretches. Head low, **<u>she</u>** walks over to **Snarl**, who purrs in pleasure and friendship. They touch faces. The tigress lies down with all four feet under **<u>her</u>**, head on the sand. As darkness falls, **Snarl** gets up, walks over to **Kulu**, and straddles **<u>her</u>**. They mate. **Snarl** bites **<u>her</u>** neck roughly and growls. **Kulu** turns **<u>her</u>** head, curls **<u>her</u>** lips, and roars right into **<u>his</u>** face. **<u>He</u>** roars back.

(*National Geographic* December 1984, p. 754)

Concerning hunting strategies, I have excerpted many examples where both male and female tigers have shared in the kill. In contrast to male lions, male tigers allow their cubs and females to feed on the kill first. (*National Geographic* December 1984). Only masculine gender markers have been excerpted (cf. E. 32.–40.):

(67) [E. 34.] "In four gigantic bounds Abu pounces on the fawn, pushes it underwater, and grabs it in **<u>his</u>** powerful jaws. **<u>He</u>** shakes it. Trotting back to shore, **<u>he</u>** disappears into the grass."

(*National Geographic* December 1984, p. 773)

Many examples of territorial behaviour have been documented. To identify their territory, males mark the trees (cf. 26.):

(68) [E. 26.] We can smell where the tiger has sprayed a bush to mark **<u>his</u>** territory. Ignoring us, **<u>he</u>** walks on. We are alongside **<u>him</u>** now, about 30 feet distant. Once or twice **<u>he</u>** glances at us and snarls. Suddenly the tiger stops in **<u>his</u>** tracks. **<u>He</u>** sees a

herd of spotted deer. **He** makes not a motion. **He** is frozen in the partial cover of a small patch of grass. As long as **he** is motionless, the deer cannot see **him**. There is no breeze, so they cannot scent **him**.

(*National Geographic* December 1984, p. 748)

I have also excerpted a variety of techniques connected with the study of these animals. Tigers have been studied in the wild or in national parks. The observations can last for years (cf. E. 23.–31.):

(69) [E. 30.] **He** is watching us and trying to catch our scent. The rest of **his** body is hidden by grass and the water in which **he** lies. Satisfied that we pose no threat, **he** lies down again. **His** stomach is grossly distended. **He** is panting so hard that waves emanate from **his** heavy body and ripple across the pond. **He** ignores us, yawning disdainfully. **His** worn yellow teeth tell us that **he** is old. The tip of **his** lower right canine is broken off. **He** is even larger than Langra. **He** walks directly toward us. In the late sun **he** shares the saja tree's shade. No more than 15 feet from our jeep **he** sits and licks the mud off **his** coat. We call **him** Saja. When **he** spies us, **he** often comes over and sits by our vehicle.

(*National Geographic* December 1984, p. 763)

The most commonly- mentioned method researchers used for observing tigers has been radio collaring (cf. E. 64., 65.):

(70) [E. 65.] Five months later **her** radio signalled no movement. **She** had vanished. Colleagues found **her** collar under the snow and **her** four-week-old cubs nearby.

(*National Geographic* July 1993, p. 43)

To sum up, personal interest plays a very important role when assigning gender to tigers. I have found one example where the tiger is being kept as a private pet. The male has his own name and is referred to by *he* (cf. E. 43.):

(71) [E. 43.] Betty Young plays with **Butterball**, one of 52 tigers at her ten-acre compound in northwest Arkansas. Young takes in tigers given up by owners unprepared for a 400-pound pet, breeds them-even trains them to use a jumbo-size litter box. After she nursed **Major Bill** from an illness, **he** took to sleeping in her bed. "**He** snores", says Young, "but I'm used to it."

(*National Geographic* December 1997, p. 30)

I have found no neuter gender markers referring to tigers. This fact can be explained by emotional factors as well as personal interest. The researchers personalise these animals, both adults and youngsters.

4.2.5.3 Leopard

(Snow Leopard, African Leopard)

Table 4.15: Gender assignment – Leopard

Pronouns	**Masculine**		**Feminine**		**Neuter**	
Personal subjective	he	34	she	21	it	3
Personal objective	him	11	her	10	it	
Possessive	his	35	her	31	its	4
Reflexive	himself	1	herself	1	itself	2
TOTAL		81		63		9

Material analysis:

The two leopard subspecies mentioned above were referred to by a total of 153 pronominal pronouns as follows:

Masculine: 81 =52.9%
Feminine: 63 =41.2%
Neuter: 9 =5.9%

Total: 152=100%

Figure 4.15: Distribution of gender – Leopard.

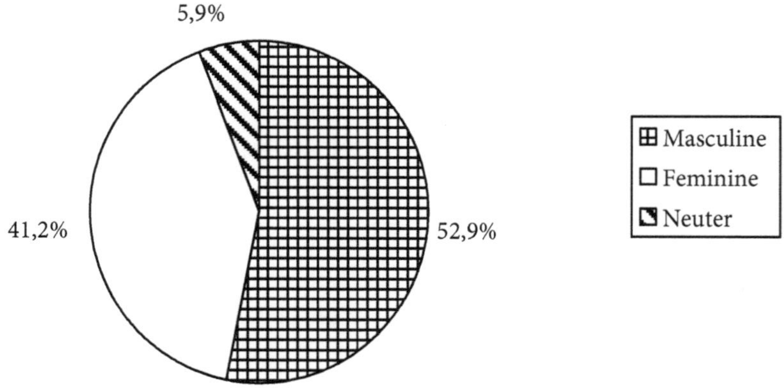

In the case of the leopard subspecies, the masculine gender dominates. The use of masculine gender markers is connected with the descriptions of hunting skills and behaviour, where the male leopard plays the main role (cf.69.–72.):

(72) [E.69.] "Pulling **his** ears back tightly against **his** head, **he** seemed to melt into the low vegetation. Lying prone, **he** was nearly invisible. Finally the leopard looked away several times, but more minutes passed before **he** made **his** move. For the first 75 yards, **he** barely lifted **his** belly off the ground as **he** crept away through short shrubs and tufts of grass. **His** body made very little motion, as if **he** had somehow located a moving sidewalk. As the animal approached the ridge on the horizon, **he** stopped, looked back, then walked out of view.

(*National Geographic* June 1986 p. 798)

Concerning the observations, researchers have studied mainly males. The animals have names, marks or numbers and they have been studied for a long time. Consequently, the researchers personalise the animals studied and their feelings are openly expressed in their language (cf.73.–76.). Of the male leopard, a naturalist wrote:

(73) [E.75.] "**He** is an animal of darkness, and even in the dark **he** travels alone." Yet over the course of 19 months one male leopard allowed me to become **his** companion. I named **him** Tjololo, which means "the one that stands alone." Years ago a veterinarian had implanted a tracking device in Tjololo to monitor **his** recovery after **he** was injured by a warthog. I was granted exclusive use of the implant's frequency, allowing me to track Tjololo's progress and gain **his** trust. Eventually I could get close enough to hear the rasping of tongue against paw as **he** groomed **himself** and the crashing of a branch as **he** marked it. Later **he** let me kneel in **his** path, eye to camera, as **he** strutted into the spotlight.

(*National Geographic* October 2001, p. 93)

In the depiction of the mating process, only feminine and masculine pronouns are used, as can be seen from the following example (cf.E.77.):

(74) [E.77.] "Mating accomplished, Tjololo springs backward off **his** partner before **she** can turn on **him**. Still, during **her** weeklong estrus **she**'ll mate with **him** hundreds of times before the pair part ways.

(*National Geographic* October 2001, p. 95)

The usage of feminine gender markers has been connected with the description of maternal care. The authors used the term "mother leopard" when describing the relationship between the female and her cubs: (cf.78.–82.)

(75) [E.78.] "They knew **the mother leopard** well. **She**'d had three new cubs the previous year, they said. The two from **her** earlier litter had gone off to establish territories of their own."

(*National Geographic* June 2008, p. 128)

The strong personal involvement of authors is obvious in the following example. The female has been numbered instead of the name and the cubs are called "babies" (cf. E. 80.):

(76) [E. 80.] "We left reluctantly, hoping that when we returned the following autumn we could recontact **4**, who would then be travelling with **her babies**. It took several months after we returned in November to confirm that **she** had two cubs. Each time **her** signal came in loud and clear, we would search intently. As we radiotracked **her** and the two growing cubs, it became evident that **she** was again utilizing **her** entire range. The cubs must have quickly developed the climbing skills to enable them to follow their **mother** whenever **she** chose to go. Months later, as Garry and I scanned the same slopes above Tillisha Stream, I spotted **Leopard 4** sauntering along the mountainside- and behind **her** were two big cubs. **Poor 4**! We wondered how **she** ever fed those two rambunctious cubs. They must bungle half **her** attempts at hunting.

(*National Geographic* June 1986, p. 101, 808)

Mostly described in plural, cubs are usually born in a litter of two to four. I have excerpted the life cycle of one female cub named Legadema, which has been observed for more than three years. The cub is viewed nearly as a pet and is always referred to as ***she*** (cf. 83.–90.):

(77) [E. 87.] "When **Legadema** was five months old, **her** mother brought **her** a live baby impala. **Her** mother guided **her** tolerantly through every step of the kill. **Her** predator skills honed, **Legadema** turned **her** attention to squirrels. Over time, **she** grew adept, killing hundreds of squirrels as well as larger prey as baby warthogs.

(*National Geographic* April 2007, p. 148)

As can be seen from the above chart, the neuter pronouns have been used in general description of the leopard's life cycle (cf. 91., 92.):

(78) [E. 91.] "While the common leopard of Asia and Africa relies on branches and leaves for concealment, the snow leopard loses **itself** among steep jumbles of stone. I imagine a leopard easing down the darkened slopes. **It** flows low to the ground. **Its** tail is almost as long and so thick and mobile it looks as if the cat is being followed by a fuzzy python. The snow leopard sometimes uses **its** tail to send signals during social encounters or to wrap partway around **itself** like a scarf when bedded down in bitter weather.

(*National Geographic* June 2008, p. 113)

Big cats- Final chart

The three species of big cats were referred to by a total of 450 pronominal pronouns as follows:

Masculine:	211=47.3%
Feminine:	223=49.8%
Neuter:	13 =2.9%
Total:	450=100%

Figure 4.16: Final chart – Big cats.

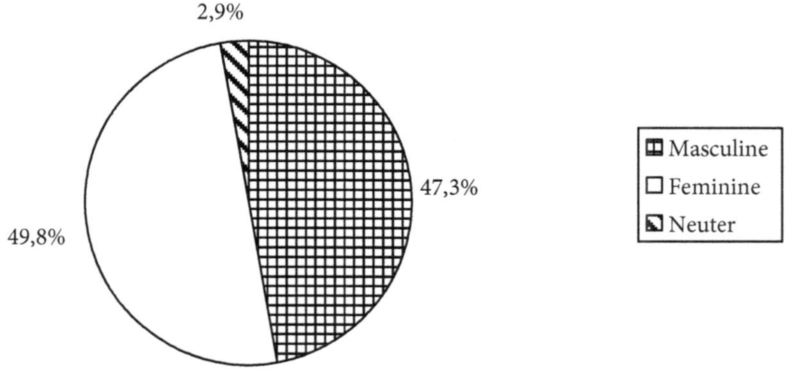

As can be seen from the data above, the usage of the feminine and masculine gender markers are balanced. The most important factor that influences gender assignment in big cats is that of emotion. The majority of excerpted examples shows both emotive factors and personal interest. The researchers and authors of the articles "feel" with these species of big cats and they talk about all of them, including youngsters, as a ***he*** or ***she***. The majority of adult animals and cubs have names, in one case a number. Researchers personalise their animals and talk about their "loved pets" or "friends". The explanation for this practice is related to the fact that these big cats belong among threatened species and are among the most recognizable and popular of the world's megafauna. The most frequently mentioned and personalised big cat in my corpus has been the tiger. The authors consider him the most charismatic animal in the world. He is the national animal of several Asian nations. (*National Geographic* December 1984)

The occurrence of neuter pronouns is very low, making up only 2.9% of all pronoun markers. The authors have used neuter in general descriptions of the life cycle of the subspecies and in two cases the lion's cub was referred to by ***it***, as mentioned above.

4.2.6 Bear

(Black Bear, Brown Bear, Grizzly Bear, Great Bear, Polar Bear, Sloth Bear, Spectacled Bear, Sun Bear, Red Panda, Giant Panda)

Table 4.16: Gender assignment – Bear

Pronouns	**Masculine**		**Feminine**		**Neuter**	
Personal subjective	he	34	she	70	it	3
Personal objective	him	11	her	22	it	
Possessive	his	24	her	54	its	5
Reflexive	himself	–	herself	4	itself	1
TOTAL		69		150		9

Material analysis:
Bears were referred to by a total of 228 pronominal pronouns as follows:

Masculine: 69 = 30.3%
Feminine: 150 = 24.2%
Neuter: 9 = 65.8%

Total: 228 = 100%

Figure 4.17: Distribution of gender – Bear.

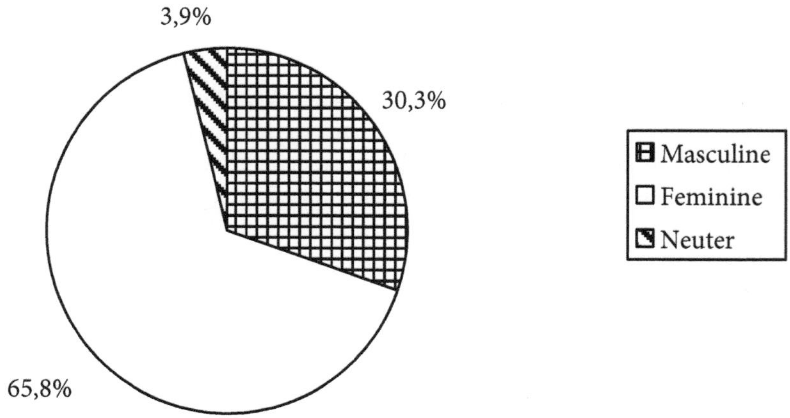

This group has been divided into two main families retrieved from the corpus, covering eight bear and two panda species.

As can be seen from the chart, feminine and masculine pronouns have been used the most frequently. In the majority of the articles explored bears have names and they are always viewed as ***he*** or ***she***. The authors have marked bears as ***he-bear*** and ***she-bear***: (cf. F.1.–F.23.):

(79) [F.10.] "The **she-bear** drops back onto **her** front feet, walks across the snow".
(*National Geographic* July 1979, p. 125)

The observed male and female bears have very often names (cf. F.6., 18.–20.):

(80) [F.6.] "I photographed **ten-year-old Bailey** at a small zoo in Grand Prairie, Texas. It is stunning to have **Bailey** walk into a small room. **His** claws, elongated from lack of use, click on the white linoleum floor as **he** tours around in that way animals have- sniffing, touching with **his** nose. And then **he** comes to me. I crouch on my haunches at eye level with **him**. **He** gets digger. I feel **his** hot breath exactly, more akin to wind. Though **he** can teak total destruction anytime **he** likes, **he** chooses to act gently."
(*National Geographic* April 1990, p. 91)

(81) [F.19.] **She** is just up here. **Squirty** eyed me and **she** walked up to greet Ben, whom **she** still seems to regard as **her** mother. Ben stroked **her** neck. **She** wasn't sure. **She** walked slowly toward me, **her** head down and teeth chomping..... When **she** was about 15 feet away, **Squirty** lunged at me and made a deep grumbling sound. Then **she** did it again.
(*National Geographic* March 2002, p. 99)

The dominance of the feminine gender can be seen in most of the quotations. They make up a total of 65.8% of all pronouns. This could be the result of the fact that majority of the examples retrieved describe the relationship between the mother and her cub or cubs. In all of the examples cited below personal interest plays a very important role. The authors thus speak about *mother bear* and *mum* when describing the relationship (cf. F.7.–21.):

(82) [F.11.] "A cub peeks around a drift as **its** twin and **mother** relax. The cubs are about three months old. **Mom** is leading **her** playful clan to the frozen bay where **she**'ll hunt.
(*National Geographic* February 2004, p. 45)

The authors used family expressions, such as *father, sister, brother* (cf. F.18.):

(83) [F.18.] "Squirty, along with **her brother** and **sister**, had been orphaned. Now grown up, with two cubs of **her** own, **she** was thriving in the wild."
(*National Geographic* March 2002, p. 95)

The majority of bear youngsters in the corpus have appeared in plural. I have found one example where a cub is referred to by *it* (cf. F.11.):

(84) [F.11.] "A cub peeks around a drift as **its** twin and mother relax. The cubs are about three months old. Mom is leading **her** playful clan to the frozen bay where **she**'ll hunt."
(*National Geographic* February 2004, p. 45)

The prevalence of plural might be explained by the fact, that the cubs of brown and polar bear species are usually born in the litters of two or three. (*National Geographic* February 2004). Consequently, I have found no example describing the relationship between the mother and her single cub.
The usage of gender markers is strongly influenced by the behaviour of these animals. Bears are typically solitary animals with the exceptions of courting individuals and mothers with their young. Consequently a number of the articles describe the life cycle of the individuals, especially concerning food and diet (cf. F. 1.,5.,14.,15.,17.,20.):

(85) [F.20.] "Ben gave Squirty some dry dog food to supplement **her** diet while **she** was nursing. For the most part, though, **she** was fending for **herself.**"
(*National Geographic* March 2002, p. 100)

In the case of **pandas**, especially the species of Giant Pandas, the feminine and masculine gender distinctions are the most frequent. The most likely explanation is that the Panda is an endangered species and critically threatened. (*National Geographic* July 2006). This is the reason for the very low occurrence of neuter. I have found only eight neuter pronouns concerning the general description of their life cycle and diet (cf. F. 44.,45.):

(86) [F.45.] "However, an animal is limited in how much **it** can eat by the size of **its** stomach. The panda has overcome these problems by harely chewing and digesting **its** food and by passing it through in only five to eight hours. Therefore, so as not to fill **itself** with useless bulk, a panda must be selective in what **it** eats.
(*National Geographic* March 1986, p. 290)

All panda bears in the corpus have names and their life cycle is described in detail (cf. F. 27.–32.,34.–37.,39.–42.):

(87) [F.37.] "Born in 1999, Hua Mei was the first surviving panda cub bred in the U.S. In 2004 **she** was recalled by China to **her** mother's birthplace-the giant panda research center in Sichuan Province's Wolong Nature Reserve-where **she** promptly got pregnant and delivered twins. Hila Mei produced a second pair of baby panda, in 2005. And **she** wasn't alone."
(*National Geographic* July 2006, p. 58)

The authors have paid great attention to the reproductive process and mating of pandas. The female panda gives birth to one or two cubs. In the case of twins, she usually abandons one of her cubs, and this cub dies soon after birth.

This behaviour is the main topic of ongoing research.(p.297, March 1986). The cubs have names and they are marked with ***he*** or ***she*** in all examples adduced (cf. F.24.,25.):

(88) [F.24.] "**He**'s got chubby cheeks. **He** naps a lot. **He** eats with **his** hands. **He** lives with **his** mother. Tai Shan is a giant panda cub, and that makes **him**, well, not your average bear. Born at 3: 41 a.m. on Saturday, July 9, 2005, at the Smithsonian National Zoological Park in Washington, D.C. **Tai Shan** is the first offspring of Tian Tian. Internet panda cams, which follow the daily activities of **Tai Shan** and **his** mom, draw an average of two million online visits a month."

(*National Geographic* July 2006, p. 48)

Researchers paid attention to their behaviour in their natural environment and in captivity. They also take attention to the special diet of pandas, which consists almost exclusively of bamboo (cf. F.27.,30.,31.,44.,45.):

(89) [F.29.] "Wei-Wei, whose name means "grand," is a middle-aged male. **His** route is typical. One January day **he** moves onto an east facing slope. Wei-Wei has rubbed the bark with the glandular area around **his** anus, depositing **his** scent. **He** remains aware of others. Sitting, **he** bends bamboo toward **his** mouth. Moving a few feet, **he** then eats some more, munching a swath through **his** domain. Once **he** ate from 3,481 stems in the course of one day. Satiated, Wei-Wei sleeps. **He** builds no nest, **he** selects no choice site; **he** merely lies on the snow. Then **he** pads on."

(*National Geographic* March 1986, p. 286)

4.2.7 Elephant

(African Elephant, Asian Elephant)

Table 4.17: Gender assignment – Elephant

Pronouns	Masculine		Feminine		Neuter	
Personal subjective	he	31	she	64	it	
Personal objective	him	17	her	31	it	
Possessive	his	43	her	98	its	
Reflexive	himself	1	herself	3	itself	
TOTAL		92		196		0

Material analysis:

The two species of elephants were referred to by a total of 288 pronominal pronouns as follows:

Masculine:	92 = 31.9%
Feminine:	196 = 68.1%
Neuter:	0 = 0%
Total:	288 = 100%

Figure 4.18: Distribution of gender – Elephant.

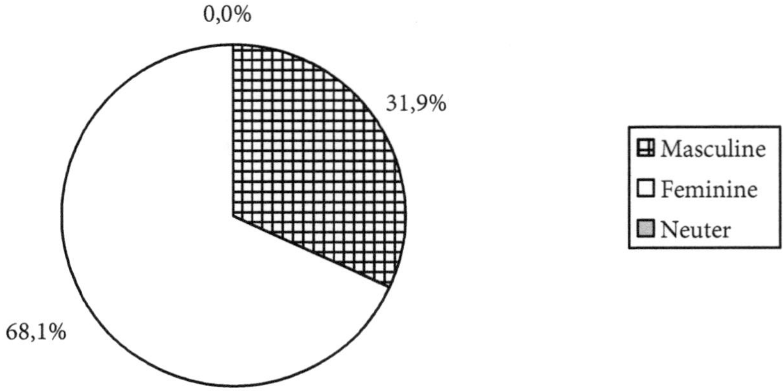

This largest land animal belongs to frequently-occurring species in National Geographic. Out of the three living species I have excerpted African Forest Elephant and Asian Elephant. A quantitative analysis has revealed that the dominant gender was feminine. The highest occurrence of feminine gender markers within the corpus could be explained by the specific life cycle of this animal. Elephants live in a structured social order. The females live in family groups made up of mothers, daughters, sisters and aunts. These families are led by the matriarch, the eldest female. The members of the groups are referred to by feminine in all excerpted examples (cf. G.30.,31., 43.,48.,52.):

(90) [G.32.] Babylon is the quintessential **matriarch**, with a hole in **<u>her</u>** left ear and a half century of experience and knowledge in **<u>her</u>** head. Yet **<u>she</u>**'s still a new **mother** as well as a **grandmother**, followed closely by **<u>her</u>** own tiny calf, and **<u>her</u>** **daughter's**, and a third. Another member of the Biblical Towns, Babel, badly lame manages to survive and stay with the family because Babylon waits for **<u>her</u>.**

(*National Geographic* September 2008, p. 52)

The authors have also paid attention to the descriptions of maternal care and breeding of the calves. The researchers personalise the mothers and calves and give them names as can be seen in the example below (cf. G.26.,33.,34.,46.,47.):

(91) [G.26.] In the city of Khon Kaen, rising from the plains of the north-eastern Thai region of Isaan, I visit a 40-year-old street elephant called **Bom** and **<u>her</u>** three-year-old **offspring**, a **big, pushy boy** named **Minimax**. They've been rented out 6Yiheir owner to two mahouts and three assistants. Although **mother** and **son** look to be in fair condition, I notice raw scars on Minimax's forehead and holes in **<u>his</u>** ears-wounds caused by an ankus, the baton with a hooked metal tip used to enforce commands.

All adult females of the close – knit groups participate in the care and protection of the young. The researchers call them "allomothers" or "full-time baby-sitters" (cf. G.32.):

(92) [G.32.] Saturn, an elderly mother **<u>herself</u>**, cares for two calves orphaned by the deaths of their mothers who were **<u>her</u>** adult daughters. At sunset along a river channel Saturn oversees the play of **<u>her</u>** own two calves and the orphans . From **<u>her</u>** behaviour alone, uniformly doting, you couldn't know which was which.
(*National Geographic* September 2008, p. 53)

The life of adult males is very different. During the breeding season a bull fights with other males and spends most of his time lingering about female herds.
In the descriptions of reproduction, males and females are always referred to by *he* and *she* (cf. G.23.–25.,G.28.):

(93) [G.24.] Males will attempt to mate with any female in estrus, but a large bull can injure a small female when mounting **<u>her</u>**-breaking one of **<u>her</u>** legs, for instance, beneath **<u>his</u>** weight. After a three-day pursuit, Leopold tries with an eight-year-old from the First Ladies and is thwarted by **<u>her</u>** mother, Mary Todd Lincoln, and other females Later, Mary Todd sanctioned **<u>her</u>** daughter's mating with a younger, smaller bull.
(*National Geographic* September 2008, p. 53)

I have excerpted many examples describing:

- life cycles and behaviours of elephants (cf. G.1.–14.,G.35.–38.)
- observation (cf. G.15.–19.,26.,27.,39.–60)
- breeding and domestication (cf. G.61.–65.)
- both legal and illegal hunting (cf. G.20.–22.). The techniques of African ivory hunters are described below:

(94) [G.22.] The man continues to slaughter **<u>him</u>** for **<u>his</u>** ivory-by poisoned arrows in Kenya. A new technique is to place poisoned fruit along **<u>his</u>** pathways.
(*National Geographic* November 1980, p. 571)

I have found no neuter gender markers referring to elephants. The most important factor that influences gender assignment in elephant species is that of emotion and personal involvement of the researchers.

4.2.8 Apes

(Chimpanzee Bonobo, Baboon, Gelada, Zanzibar Red Colobus Monkey, Proboscis monkey, Patas monkey, Gorilla, Orangutan)

Table 4.18: Gender assignment – Apes

Pronouns	Masculine		Feminine		Neuter	
Personal subjective	he	59	She	49	it	1
Personal objective	him	21	Her	25	it	2
Possessive	his	68	Her	68	its	6
Reflexive	himself	2	herself	2	itself	
TOTAL		150		144		9

<u>Material analysis:</u>

The eight species of apes were referred to by a total of 303 pronominal pronouns as follows:

Masculine: 150 = 49.5%
Feminine: 144 = 47.5%
Neuter: 6 = 3%

Total: 303 = 100%

Figure 4.19: Distribution of gender – Apes.

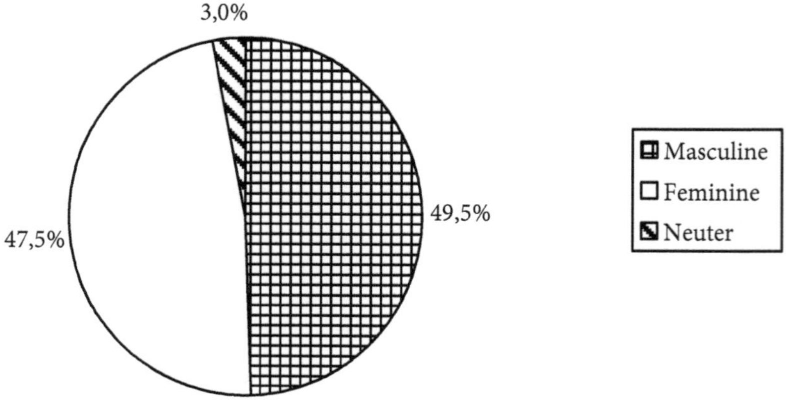

Out of eight species of apes, three of them were prevalent. The majority of articles described mainly the **Gorilla, Orangutan** and **Chimpanzee** species.

The analysis has revealed that both feminine and masculine genders are dominant. The neuter is marginal. I have excerpted only eight neuter pronouns, all of them concerning juveniles. Even if the researchers talk about the juveniles as babies, I have found only neuter references (cf. H. 70.–74.):

(95) [H. 70.] "A four-week-old baby rests with **its** mother in the relative peace in sabah."
(*National Geographic* August 2002, p. 115)

The majority of examples describe the behaviour and social structure of different species. These studies have shown that in their natural environments, different apes display varying social roles. The most frequent topics are descriptions of families (cf. H. 9.–23., H. 34., H. 51.–57.):

(96) [H. 12.] "Banging the cans to intimidate larger males, **he** bluffed **his** way to the top. Passion killed and ate Gilka's infant, Otta, sharing the flesh with **her** daughter, Pom. Figan's status as the alpha male had been bolstered by **his** close relationship with **his** brother Faben. When Faben disappeared, Figan had a hard time fending off challengers by **himself**- although **he** managed to reign for ten years. After three-year-old Mel's mother died of pneumonia, **he** was adopted by an adolescent male."
(*National Geographic* December 1995, p. 110, 111)

The apes perform well on a wide range of cognitive tests. The corpus includes descriptions of long-term observations (cf. H. 6., 7., 8., H35.–46., H58.–69.) and studies in animal language acquisition, which have been completed with great apes (cf. H. 4., 5.):

(97) [H. 6.] "This takes years of closely following a silverback and **his** family-essentially living with the gorillas.
(*National Geographic* January 2008, p. 94)

(98) [H. 4.] "There was no doubt in my mind that Sugito could learn sign language. **He** already was capable of simple symbolic communication-for instance, handing me the key to the storeroom when **he** wanted food. But, unfortunately, although Sugito stopped killing infants, **he** turned increasingly dangerous and ugly as **he** grew older. **He** bit a few workmen and repeatedly raided camp. Gary could not get near **him**, let alone teach **him** anything.
Eventually Rod and Gary caught Sugito after one of **his** most destructive raids and released **him** in the forest a considerable distance from our camp. Sugito was extraordinarily clever, and I have no doubt **he** survived quite well on **his** own."
(*National Geographic* June 1980, p. 846)

The close relationship between the apes and the authors is evident. The animals are divided into study groups and the majority of the individuals have names, nicknames or marks, including the youngsters (cf. H.71.):

(99) [H.71.] "Families are headed by females like Theodora, here with **her** black infant, **its** playmate, and a male friend, Pinocchio. Closest to me huddled Thea and baby Thistle, with oldest daughter Theodora and **her** infant, Teleki."
(*National Geographic* November 1987, p. 677)

In the majority of these research studies a very strong relationship is obvious between the mother and her juveniles.

To sum up, the most significant factor that influences gender assignment in apes is that of emotion. All the animals are referred to with personal interest. The researchers personalise the animals and talk of them as ***he*** and ***she***. Emotions play an important role especially when the articles concern a disease or victims of poachers as shown in the two following examples (cf. H 24.–27.,47.,48.):

(100) [H.27.] "Weeks later I hear that a veterinarian based in Rwanda has darted the youngster, saved **his** foot, and christened **him** Bahati- Swahili for "luck."
(*National Geographic* October 1995, p. 80)

(101) [H.24.] "Another victim of poachers, Uncle Bert left behind five offspring, including baby daughter Frito. In the subsequent power play, common after the death of a group leader, Beetsme, a contending silverback, killed Frito. **He** acted instinctively, the author believes, ensuring the dominance of **his** future bloodline."
(*National Geographic* April 1981, p. 519)

4.2.9 Final chart and material analysis

All analysed species taken from zoological articles of the *National Geographic* have been referred to by a total of 2753 pronominal pronouns as follows:

Masculine:	906 = 32.9%
Feminine:	1284 = 46.6%
Neuter:	563 = 20.5%
Total:	2753 = 100%

Figure 4.20: Final chart – National Geographic Material analysis.

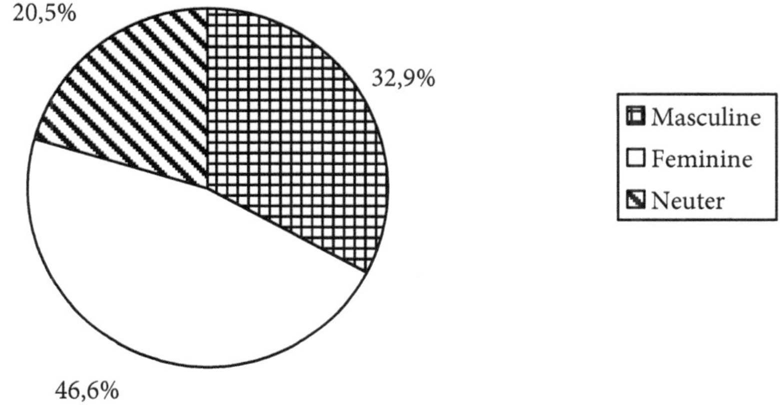

Overall, the final counts of gender markers in the articles under study show that the feminine gender is the most frequent when referring to all animal species explored. The quantitative analysis of the data reveals the prevalence of the feminine gender in all animal species in question. The "lowest" animals with gender distinctions are insect species (77.5% of all pronouns are feminine). The animals with the highest occurrence of **feminine** pronouns are ants (86 %), elephants (68.1%), bears (65.8%), lions (58.9%), tigers (53.2%), reptiles and amphibians (48.7%) and whales (45.6%). The animals are assigned to the feminine gender predominantly in situations referring to maternal care and reproduction.

The animals with the highest occurrence of **masculine** pronouns are seahorses (74.4%), exotic birds (56.7%), birds of prey (53.1%), leopards (52.9%) and apes (49.5%). The predominance of masculine in these species is probably the result of extensive descriptions of behaviours, life cycle and reproduction processes.

Tiger, elephant and seahorse species have been referred to only by feminine and masculine. Neuter has been totally absent. A possible explanation of the unusual gender assignment might be strong personal involvement with threatened tiger and elephant species and the unique reproductive process in seahorse species.

The occurrence of **neuter** in the analysed material is surprisingly marginal. It covers only 22.8% of all pronoun markers. The use of neuter is connected with the general descriptions of the life cycle without any personal involvement of the authors. Neuter has been most frequent when referring to marine microfauna (75%), tropical fishes (73.9%) and shark species (53.7%). The possible explanation of the neuter gender prevalence in these species might be the wide variety of marine microfauna and fish species. The authors have had no need to indicate the gender. General descriptions of life cycle have prevailed. In the case of shark species, the main reason for using predominantly neuter could be little information obtained in recent studies concerning its behaviour and reproduction.

Concluding from the material analysed, the assignment of gender to animal nouns in *National Geographic* articles does not fully follow the rules of a semantic assignment system. The majority of animal species are assigned to genders according to their biological sex, but other important factors also influence the gender marking. Sex differences are indicated by a range of gender markers where they are felt to be relevant. The choice of the pronoun depends also on the social roles of the animals and their status in "the animal kingdom". The animal species are not necessarily parallel to biological classification, so lower animals may often be viewed by authors as higher animals. Apart from the grammatical functions of the relevant pronouns , the pronouns imply a sociological function. Based on the corpus analysed, the prevailing theory that *the pronoun* **it** *is used to refer to animals predominantly in scientific or non-literary contexts where there is no need to express personal or social attitudes* has been shown to be invalid.

4.3 The analysis of gender markers in animals based on *Journal of Zoology*

The aim of this chapter is to collect, compare and analyse animal species retrieved from scientific zoological papers. In all of them the focus is on gender of animal nouns.

4.3.1 Journal of Zoology

The field of zoology is extremely broad and covers a number of disciplines. There are many specialist journals currently available. I have chosen the *Journal of Zoology* for a number of reasons. The first is the very breadth of research covered. Second, the journal has published original research papers and reviews within the whole field of zoology for about 170 years. Over that time, it has built up an impressive reputation, receiving contributions from zoologists around the world and achieving an ever-increasing international readership. This journal also reflects an enormous variety in the field. Even in volume 271, January 2007 we can find topics that could be published in specialist ecological, behavioural or evolution journals. Yet there are other papers that cut across anatomy, systematics, genetics, phylogenetics and developmental biology. This journal is published monthly, 12 times a year as three volumes. It is published on behalf of The Zoological Society of London.

The corpus for the first analysis includes **276** research papers taken from the volumes 268–274 **(2006–2008).** All analysed papers were in English and published in London in 2006–2008.

I recorded **217** animal nouns and their gender references – all personal, possessive and reflexive pronouns that referred to animal nouns in the corpus of the papers. Many animals in the corpus were without gender references. The authors have used mostly specific biological terms, marks or plural, when referring to animal species. I have listed only animals with gender references. The focus is primarily on the occurrence of masculine and feminine pronouns. The species assigned to by masculine and feminine are analysed in more detail in the following chapter. All examples retrieved can be found in the appendices.

The next table lists all animals in alphabetical order with all gender markers that were identified:

Table 4.19: Gender assignment in animals taken from Journal of Zoology

Animal	**Masculine**	**Feminine**	**Neuter**
AFRICAN ELEPHANT			1
AFRICAN GROUNS SQUIRREL			1
AFRICAN LION			1
AFRICAN TOAD			2
AFRICAN WILD DOG			1
ALBATROSS	**16**	**12**	12
ALGERIAN HEDGEHOG			3
ALPINE LYNX			4
AMERICAN PITVIPER			3
AMERICAN SQUIRREL			1
ANT BUG			5
ANT PHEIDOLE			1
ANURAN			1
ARMADILLO LIZARD			2
ARMORED SHREW			26
ASIAN ELEPHANT			1
ATLANTIC SALMON			5

Animal	**Masculine**	**Feminine**	**Neuter**
AUSTRALIAN SLEEPY LIZARD			3
BADGER			2
BALL PYTHON			1
BANDICOOT RAT			1
BEAVER			2
BENT-WINGED BAT			10
BIG BROWN BAT			2
BINTURONG			2
BEAR	**3**	**6**	10
BLACK HOWLER MONKEY			2
BLACK-BACKED JACKAL			1
BLACK-FACED IMPALA			1
BOBCAT			7
BONELLI'S EAGLE			2
BOTTLENOSE DOLPHIN			1
BRENT GOOSE			1
BROWN HARE			2
BRUSHTAIL POSSUM			3
BRUSH-TAILED MULGARA			1
BUTTERFLY			1
CARIBOU			1
CHEETAH			1
CHIMPANZEE			2
CICHLID			5
CIVET			1
COBRA			4
COLONIAL BAT			1
COTTONMOUTH			2
COUGAR (PUMA)			2
COYOT			3
CRAB	3	12	5
CRAB-EATING FOX			1
CRAYFISH		2	18

Animal	**Masculine**	**Feminine**	**Neuter**
DARWIN'S FOX	10	2	10
DAUBENTON'S BAT			10
DESERT COBRA			2
DIVING BEETLE			4
DOMESTIC CAT			3
DORMOUSE			5
DRAGON LIZARD		6	20
DUNGENESS CRAB			5
DWARF DEER			4
EGG-EATING SNAKE			1
ELEPHANT			1
EMBALLONURID BAT			9
EURASIAN BADGER			1
EURASIAN LYNX	3	9	7
EURASIAN OTTER			1
EUROASIAN HUTHATCH			2
EUROPEAN BADGER			2
EUROPEAN FREE-TAILED BAT			7
EUROPEAN HARE			2
EUROPEAN HEDGEHOG			1
FIELD CRICKET	10	6	6
FLORIDA MANATEE			1
FORREST BUFFALO			3
FOUR-HORNED ANTELOPE			2
FREE- RANGING DOG			1
FRESHWATER CRAYFISH			1
FRESHWATER SNAIL			1
FRESHWATER TURTLE		30	11
FROG			2
FRUIT BAT	1		3
FUR SEAL		1	6
GARTER SNAKE			1
GAZELLE			2

Animal	Masculine	Feminine	Neuter
GECKO			1
GEKKONID LIZARD		2	
GENET			1
GIANT ANTEATER			11
GIANT ANTEATER			2
GOBIID FISH			1
GOLDEN MOLE			6
GRASS SNAKE			1
GRASSHOPPER			6
GREEN FROG			1
GREY CANGAROO			1
GREY- CROWNED BABBLER			1
GREY KANGAROO			1
GREY WOLF	1		1
GROUND SQUIRREL			1
GUINEA PIG			1
HAMLET			1
HAMSTER			1
HARBOUR SEAL			1
HEDGEHOG			1
HERRING GULL			1
HONEY-BZZARD			2
HOPPING MOUSE			2
HORSESHOE BAT			3
HOUSE MOUSE			2
HUEMUL			1
HUMPBACK WHALE			3
HYAENA			7
IBERIAN SPADEOOT TOAD			2
IMPALA			2
INDIAN TROPICAL BAT			1
INDIAN WOLF		3	1
JACKRABBIT			2

Animal	Masculine	Feminine	Neuter
JAGUAR			2
JAPANESE MACAQUE			1
KARNEL BLUE BUTTERFLY			1
KILLER WHALE			1
KIWI			2
KOMODO DRAGON			5
LAYSAN ALBATROSS			1
LEAF-TAILED GECKO			1
LEMUR OF MADAGASKAR			2
LEOPARD			1
LION			5
LONG- FINGERED BAT			1
LONG-EARED BAT	1	9	1
LONG-LIVED TURTLE		2	
MACAQUE			1
MANED WOLF		17	3
MARCH HARRIER			5
MOLE CRICKET			2
MOLE RAT			2
MONGOLIAN GAZELLE			1
MONKEY			3
MORELET'S CROCODILE			2
MOUNTAIN BRUSHTAIL BOBUCK			1
MOUNTAIN YELLOW-LEGGED FROG			6
MOUSE			1
MUSKOX			1
NATIVE CENTIPEDE			2
NORWAY RAT			1
NORWAY RAT			2
OCELOT			2
OTTER			2

Animal	Masculine	Feminine	Neuter
PAINTED STORK			5
PAMPAS FOX			1
PANDA			6
PANDALID SCHRIMP			1
PARROT			4
PECCARY			1
PENGUIN			2
PERSIAN WILD ASS			1
PINE MARTEN			1
POLAR BEAR			2
POND TURTLE			2
POSSUM	4	4	2
PRAYING MANTIS	4	7	6
PRZEWALSKI HORSE	2	1	5
PUMA			3
PYGMY LIZARD			1
PYGMY POSSUM			1
RACOON			2
RACOON DOG			1
RATTLESNAKE		4	1
RED DEER			3
RED FOX			1
RED SWAMP CRAYFISH			4
RIDLEY TURTLE			1
ROE DEER			1
SALAMANDER			1
SAVANNA ELEPHANT			1
SCANDINAVIAN WOLF			1
SCHREIBER'S BAT			1
SEA LION			2
SEA TURTLE			6
SHEEP			1
SHORT- TAILED FRUIT BAT			1

Animal	Masculine	Feminine	Neuter
SIKA DEER			1
SLOTH BEAR			5
SPIDER			14
SPRING HARE			1
STAG BEETLE			2
STONE MARTEN			2
STONY CREEK FROG			1
SUN BEAR			11
SWIMBLADDER FISH			1
TAMMAR WALLABY			4
TASMANIAN DEVIL			2
TERMITE			2
TERRAPIN			1
TERRESTRIAL SALAMANDER			1
TIBET YAK			2
TIBETAN FROG			2
TIBETAN GAZELLE			2
TIGER			2
TREEFROG			3
TRUE VIPER			2
TURTLE			1
VATER VOLE			1
WALL LIZARD			1
WANDERING ALBATROSS			6
WATER DRAGON			1
WATER WOLE			3
WATERSNAKE		1	1
WEASEL			2
WESTERN GORILLAS			4
WHITE-BREASTED HEDGEHOG			2
WILD DOG	5	6	2
WILD MEERKAT			1
WOLF SPIDER		9	20

Animal	Masculine	Feminine	Neuter
WOOD MOUSE			1
WOODPECKER			1
YELLOW AND BANDED MONGOOSE	1	5	7
ZEBU			2
Total	**64**	**156**	**624**

Material analysis

The following chart shows the proportion of masculine, feminine and neuter references based on their occurrence in the analysed papers.

217 species of animals were referred to by 844 pronouns as follows:

Masculine: 64 = 7.6%
Feminine: 156= 18.5%
Neuter: 624= 73.9%

Total: 844= 100%

Figure 4.21: Distribution of gender – animal species taken from Journal of Zoology.

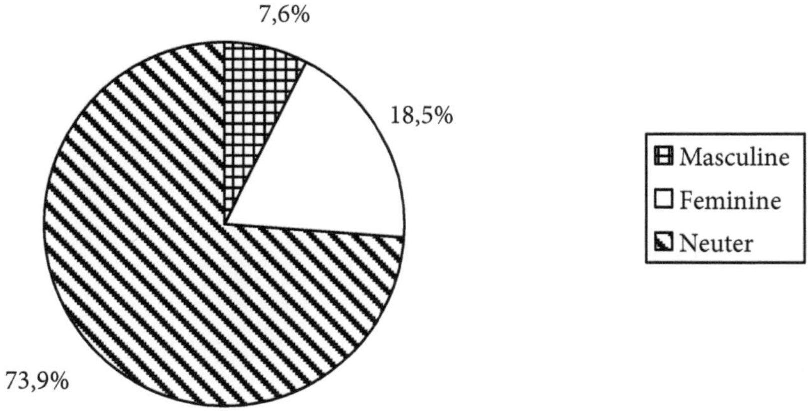

The results, as can be seen from this table and the chart, show the preference for neuter when analysing all the species. The preference for using neuter in scientific papers is evident, as there is no need to express the gender of each animal analysed. Instead of pronoun markers researchers use the name of the studied animal, its mark or a specific scientific term. The studied individuals are often referred to by plural forms.

All 217 animal species mentioned in the above table contain gender markers. 73.9% of animals have been assigned only to neuter. A total of 25 of studied species include masculine or feminine gender references. On the other hand, the feminine gender assignment is surprisingly high. One possible explanation of this could be that the animals analysed are mostly females with their young, a category most interesting for researchers. To sum up, in all the selected examples the assignment of gender to animal nouns is primarily based on their natural biological sex but another important factor that influences gender assignment within this corpus is that of **emotion** and **personal involvement** of the researchers.

4.4 The comparison of gender markers in animals included in the corpora of *National Geographic* and *Journal of Zoology*

The aim of this analysis is to ascertain the differences in gender assignment in animals in two different sources- the non-fiction literature as represented by *National Geographic,* and the research zoological papers included in *Journal of Zoology*.

The tables below compare animal species with masculine and feminine pronouns taken from *National Geographic* corpus and the same animal species taken from the *Journal of Zoology.*

I have found only 16 identical species of animals within the two corpora under study. I have analysed and compared their masculine and feminine gender references retrieved from both corpora. The following is the list of identical species under study:

mantis, cricket, spider, crab, turtle, lizard, snake, bat, albatross, seal, wild dog, wolf, fox, lynx, bear and horse.

I have arranged the analysed animals according to the biological classification, starting with the lowest insect and finishing with big mammals. There are no comparisons for crayfish, possum and mongoose, which were retrieved from the *Journal of Zoology*. I have found no appropriate animal species in the *National Geographic* articles. Consequently, these three species have not been included in this part of the research and in the final analysis.

Each animal species and the pronouns referring to it are recorded in a separate table. The classes and species of the animals found in both sources are given in brackets above each table. The tables cover gender markers taken from both corpora. They are followed by examples taken from both sources, printed in italics. I have highlighted in quotations both pronouns and names referring to the analysed animals to make the comparison clearer. Each animal species has a pair of the final comparative charts.

4.4.1 Praying Mantis

genus Ciulfina- Journal of Zoology

Table 4.20: Comparative gender assignment – Praying Mantis

Pronouns	**Masculine**			**Feminine**			**Neuter**		
		NG	JZ		NG	JZ		NG	JZ
Personal subjective	he		2	she	4	2	it	3	4
Personal objective	him	1	1	her	1	5	it		
Possessive	his	5	1	her	7		its	5	3
Reflexive	himself			herself			itself	1	
TOTAL		**6**	**4**		**12**	**7**		**9**	**6**

<u>Material analysis:</u>

Figure 4.22: Comparative charts – Praying Mantis.

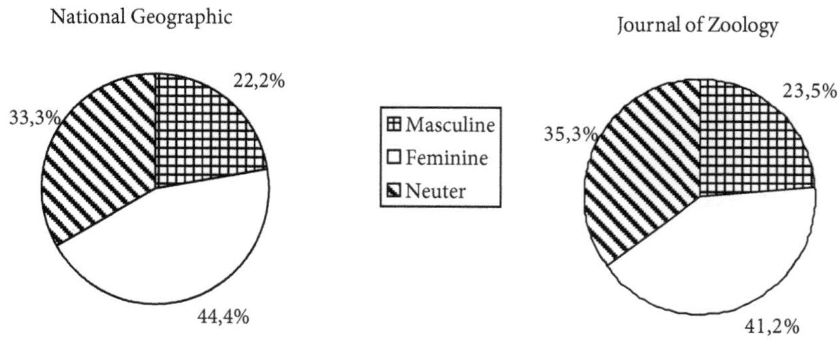

Chyba! Chybné propojení.

As can be seen from the comparative charts, the occurrence of all three genders in praying mantis is remarkably similar in both sources. The majority of the examples retrieved refer to the very specific reproductive process of this insect, which thus requires finer gender distinction. In both sources males and females are assigned predominantly to ***he*** and ***she***. The dominance of feminine gender markers can be explained by the female's main role in the reproduction of praying mantis. The following examples taken from both sources show the gender assignment in the description of observation:

(97) [A.65.] As mating proceeds, the female clutches the male around the neck and begins to feed on **his** head. Though now decapitated, the male has a nervous system that enables **him** to continue mating. Some scientists have suggested that the male thus benefits **his** offspring by providing protein for the female during egg production.

(*National Geographic* p. 274, vol. 165, 1984.)

(98) [J. 1.] If the male faced the front of the female, **he** would dart quickly around the tree trunk and move into position behind **her**.....Once the male' s head was within c. 5–l0mm of the female's abdomen, **he** would leap onto the female's back, gripping **her** thorax or midlegs with **his** forelegs and insert **his** genitalia into **her** genital opening. If the female became aware of the male before a mating attempt, **she** would rapidly turn to face **him** or run away........ ..
......the female used a hindleg to twist **her** abdomen around, forming a 180 U-shape, and used **her** mouthparts to remove the spermatophore and consume it. **She** then groomed **her** opening.

(*Journal of Zoology* vol. 271, 2007, p. 257)

4.4.2 Cricket

(Corn cricket, Tree weta, Mountain weta, Grasshopper)-*National Geographic* Alpine bushcricket (*Anonconotus*), Field cricket *(Gryllus bimaculatus)* – *Journal of Zoology*

Table 4.21: Comparative gender assignment – Cricket

Pronouns	Masculine			Feminine			Neuter		
		NG	JZ		NG	JZ		NG	JZ
Personal subjective	he		1	she		2	it		6
Personal objective	him		1	her		3	it	1	
Possessive	his		7	her	1	1	its	4	
Reflexive	himself		1	herself			itself		
TOTAL			**10**		**1**	**6**		**5**	**6**

Material analysis:

Figure 4.23: Comparative charts – Cricket.

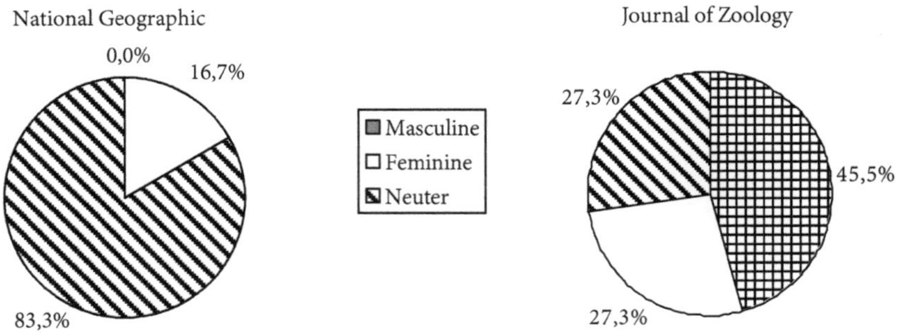

Chyba! Chybné propojení.

Gender distinction in this species of insects is completely different when comparing the two sources. Even though the number of cricket species explored was higher in the *National Geographic* corpus, I have surprisingly not found any masculine, and only one feminine pronoun referring to cricket. All the retrieved examples of masculine and feminine pronouns refer to observation and life cycle (cf.N.1.–N.4.):

(99) [N.1.] About as big as a house mouse, this weta weighs in at two thirds of **her** species' record 2, 5 ounces.

(*National Geographic* November 1991, p. 101)

In research papers of *Journal of Zoology* the examples refer to courtship and reproduction of the cricket species. The number of masculine and feminine markers is very high, males and females are predominantly referred to as ***he*** and ***she*** (cf.J.2.–J.5.):

(100) [J.4.] During courtship, male field crickets often posture and antennae the female, laying or lashing antennae over **her**.

(*Journal of Zoology* vol. 268, p. 341)

(101) [J.2.] The male would then rapidly re- position **himself**, moving **his** body underneath the female's abdomen, with **his** head facing **her** ovipositor. **He** would grip the ovipositor with **his** four front tarsi…

(*Journal of Zoology* vol. 276, p. 315)

4.4.3 Spider

Portia fimbriata, Deinopis, Coneweb spider, Tarantula, Sun spider-*National Geographic*

Huntsman Spider (*Araneae: Sparassidae*), Spitting spider *(Scydodes pallida)*, Funnelweb spider *(Hadronyche)*,Jumping spider *(Araneae: Salticidae)-Journal of Zoology*

Table 4.22: Comparative gender assignment – Spider

Pronouns	Masculine			Feminine			Neuter		
		NG	JZ		NG	JZ		NG	JZ
Personal subjective	he			she	8	4	it	24	8
Personal objective	him	2		her	2		it		1
Possessive	his	3		her	9	5	its	25	9
Reflexive	himself			herself			itself	3	2
TOTAL		**5**			**19**	**9**		**50**	**20**

<u>**Material analysis:**</u>

Figure 4.24: Comparative charts – Spider.

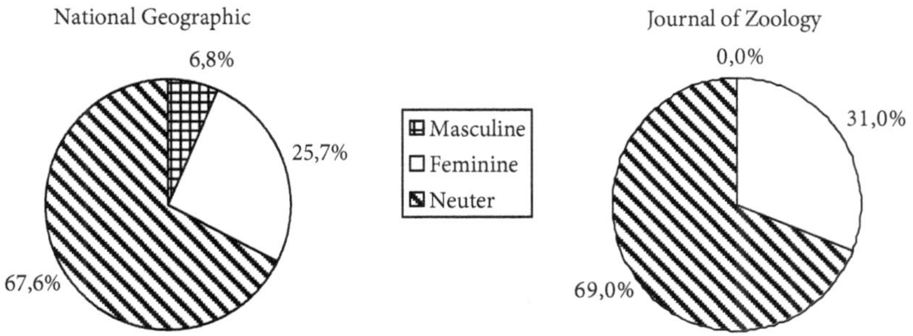

Chyba! Chybné propojení.

In the case of spider species the prevalence of neuter is evident in both sources. In the *National Geographic* corpus I have found five spider species. The authors have used predominantly neuter. Masculine and feminine references occur particularly in descriptions of life cycles and obtaining food (cf.N.5.–8.):

(102) [N.6.] **She** eats **her** booty one egg at a time, puding each into **her** mouth with **her** palps.
.....Portia dropped on **her** own silk line alongside the web. Then Portia began to swing toward **her** victim...until **she** made a kill.
(*National Geographic.*, vol. 190, 1996, p. 114)

Four species taken from *Journal of Zoology* have been referred to mainly by neuter. The rest of gender markers are feminine. They refer to maternal care. That is why the female and her offspring are mentioned in the majority of examples (cf. J.6.–8.). The researchers use the terms *mother, young, offspring, spiderlings,* as can be seen in the following example:

(103) [J.6.] This form of **maternal 'care'**, otherwise referred to as 'tolerant' behaviour, comprises passive care behaviours. For example, **a mother** may protect **her young** from predators as a passive consequence of **her** presence, and food items in the form of discarded carcasses that **she** has previously fed upon may be available for **her offspring** to 'pick over.......... Spiderlings (p. 235)
(*Journal of Zoology* vol. 271, 2007, p. 233)

4.4.4 Crab

Spider crab, Red crab, Horseshoe crab, Blue crab, Mud crab-*National Geographic*

Hermit crab –*Journal of Zoology*

Table 4.23: Comparative gender assignment – Crab

Pronouns	Masculine			Feminine			Neuter		
		NG	JZ		NG	JZ		NG	JZ
Personal subjective	he		2	she	7		it	6	5
Personal objective	him	2		her	3	4	it		
Possessive	his	4	1	her	14	8	its	14	
Reflexive	himself			herself			itself	2	
TOTAL		**6**	**3**		**24**	**12**		**22**	**5**

Material analysis

Figure 4.25: Comparative charts – Crab.

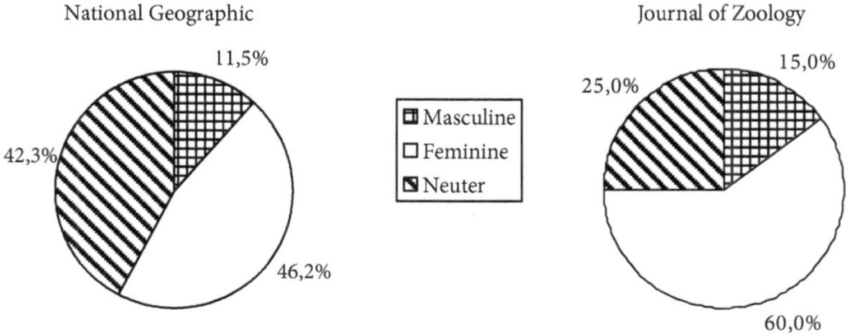

Chyba! Chybné propojení.

Five crab species taken from *National Geographic* corpus have been referred to as feminine in the majority of examples. Thematically, the articles dealt with reproduction, maternity care, egg laying and nest building (cf. N.9.–12.). The following example describes mating in crabs, therefore the pair are viewed as ***he*** and ***she***, respectively:

(104) [N.9.] In summer and early fall the female, a "**she crab**" to watermen, prepares to molt for the very last time. **She** rocks from side to side, waving **her** claws. The male, a "**jimmy**," then cradles **her** with **his** walking legs and takes **her** with **him** in search of a secluded place out of harm's way. The female backs out of **her** shell and is ready-for the only time in **her** life-to mate. **She** nestles beneath the male in a copulatory embrace that may last 5 to 12 hours. Once the mating is complete, the male continues to cradle the female, now called a sook, for as long as 48 hours until **her** shell has solidified. **She** is also known as a sponge crab for the soft mass of eggs on **her** underside. If the intimate exchange takes place in the fall, the female hurries off to find a burying ground. There **she** will snuggle into the sediment for the winter, storing the male's sperm packet until spawning time the following spring. The male finds **his** own place to burrow, perhaps returning in the spring to the piling life that sustains **him**.
(*National Geographic* July 1992, p. 114)

The neuter is used mainly for the description of life cycle (cf. N. 13.–14.).

Very similar rules of gender assignment have been found to apply in the research papers. Gender distinctions occur only in one of the species explored. The authors mark males and females as ***he*** and ***she***. The focus is on mating:

(105) [J.9.] Mating in these animals takes place when the male grabs the female's shell or body, inducing **her** to come out of the structure. The female then shows **her** disposition to mate by exposing **her** cephalothorax.

...they may still be controlled by **her** as the gametes are kept within **her** shell. When male finds a female, **he** drags **her** by grasping **her** shell using **his** left minor cheliped. The process, which is that male behaviour intends to exhaust the female, weakening **her** and thus facilitating **her** exit from the shell...

(*Journal of Zoology* vol. 270, p. 596)

4.4.5 Turtle

Alligator Snapping Turtle, Green Sea Turtle, Hawksbill Sea Turtle, Leatherbask Sea Turtle, Loggerhead Sea Turtle, Olive Ridley Sea Turtle, Coahuilan box turtle -*National Geographic*

Sea Turtle (*Caretta caretta*), Green turtle *(Chelonia mydas)*, Pond turtle *(Emys orbicularis)-Journal of Zoology*

Table 4.24: Comparative gender assignment – Turtle

Pronouns	Masculine			Feminine			Neuter		
		NG	JZ		NG	JZ		NG	JZ
Personal subjective	he	3		she	8	23	it	17	6
Personal objective	him	4		her	7		it	9	
Possessive	his	6		her	12	9	its	10	3
Reflexive	himself			herself	1		itself		2
TOTAL		**13**			**28**	**32**		**33**	**11**

<u>Material analysis</u>

Figure 4.26: Comparative charts – Turtle.

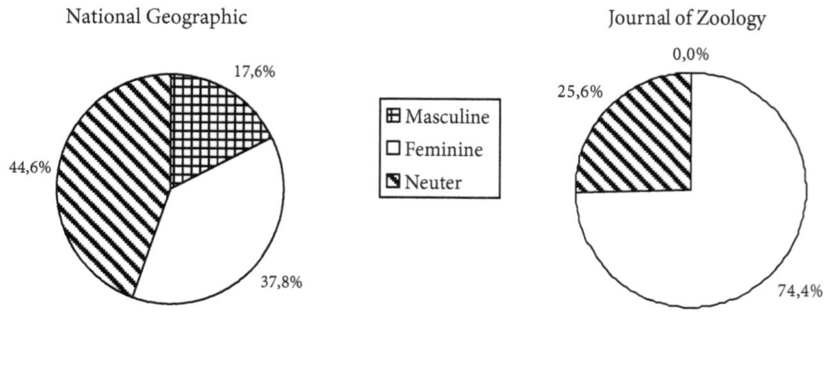

Seven turtle species are mainly referred to by feminine and neuter. The feminine is used in the descriptions of maternal care and egg laying:

(106) [C.32.] **She** digs a nest and lays about 150 eggs. **Her** return to water is just as swift. An hour earlier **she** had started to come up, but something wasn't quite right, and **she** left without laying. Now, heavy with eggs**, she** was watching, listening, waiting.

(*National Geographic* vol. 185,1994, p. 115)

Both feminine and masculine are used in the description of mating (C.29.–34.).

In *Journal of Zoology* papers the feminine gender dominates. All retrieved examples refer to observation. The wandering female is therefore viewed as ***she*** (cf. J.10.–15.):

(107) [J.12.] When **she** was located in coastal waters of Kyushu, most of the seafloor depths for **her** positions were 50–150 m. During these periods, **she** spent most of the time at 0–150 m depths. Thus, **she** likely reached the bottom in some dives during these periods. **She** did not show a diel pattern in diving behaviour until **she** moved to a resident foraging area at the continental shelf of the East China Sea. During the residence at the shelf edge, **she** spent most of the time at 100–150m depths during the day, while **she** almost exclusively used shallow depths (0–25 m) at night. Because most of the seafloor depths where **she** resided were 120-150m, **she** likely dived to the bottom during a day. **Her** mean dive durations were not different…….

(*Journal of Zoology* vol. 273, 2007 p. 50)

4.4.6 Lizard

Angolosaurus lizard, Draco lizard, Green Basilisk lizard, Short-Horned lizard, Water- skimming lizard, Gecko, Chameleon, Tuatara, Agama, Brandt's lizard, Iguana, Mulgara, Thorny Devil-*National Geographic*

Dragon Lizard *(Rankinia diemensis)*,*Gallotia galloti*, Eyed lizard *(Lacerta lepida)*, Gekkonid lizard *(Hoplodactylus maculatus)*, Sleepy lizard *(Tiliqua rugosa)*,Tropical lizard *(Carlia rubrigularis)*, Lizard *(Sceloporus malachiticus)*-*Journal of Zoology*

Table 4.25: Comparative gender assignment – Lizard

Pronouns	Masculine			Feminine			Neuter		
		NG	JZ		NG	JZ		NG	JZ
Personal subjective	he	1		she	2	4	it	8	16
Personal objective	him			her			it		1
Possessive	his			her	5	4	its	23	8
Reflexive	himself			herself			itself	1	
TOTAL		**1**			**7**	**8**		**32**	**25**

Material analysis

Figure 4.27: Comparative charts – Lizard.

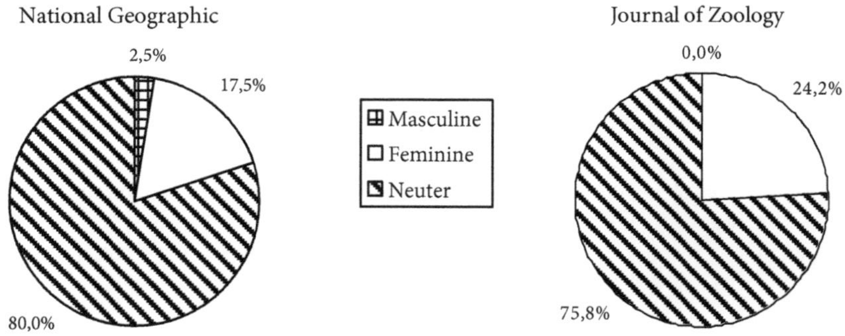

The dominant gender in both sources is neuter. The occurrence of other gender markers is very low.

When referring to thirteen species taken from *National Geographic*, feminine and masculine is used when describing observation (cf. C. 12.–13.)

Three lizard species taken from *Journal of Zoology* are referred to by feminine in the descriptions of reproduction and the study of egg laying process (cf. J. 16., 17.):

(108) [J. 16.] The data revealed that for a mature female, who first averages c. 4 eggs per clutch, for every additional egg **she** must first gain c. 2.4 g in body mass. As the average female mass in this study is c. 10 g, this means that **she** must increase **her** body size by almost a quarter of **her** body mass in order to increase **her** reproductive output

(*Journal of Zoology* vol. 273, 2007, p. 270)

4.4.7 Snake

Adder, Eastern diamondback Rattlesnake, Anaconda, King cobra- ***National Geographic***

Rattlesnake ***(Crotalus cerastes),*** **Rainbow boa** ***(Epicrates cenchria maurus),*** **Watersnake** ***(Nerodia sipedon),*** **Green python** ***(Morelia viridis),*** **Garter snake** ***(Thamnophis sirtalis)-Journal of Zoology***

Table 4.26: Comparative gender assignment – Snake

Pronouns	**Masculine**			**Feminine**			**Neuter**		
		NG	**JZ**		**NG**	**JZ**		**NG**	**JZ**
Personal subjective	he	1		she	5		it	10	2
Personal objective	him			her	4	1	it		
Possessive	his	1		her	6	4	its	7	2
Reflexive	himself			herself			itself		
TOTAL		**2**			**15**	**5**		**17**	**4**

<u>**Material analysis**</u>

Figure 4.28: Comparative charts – Snake.

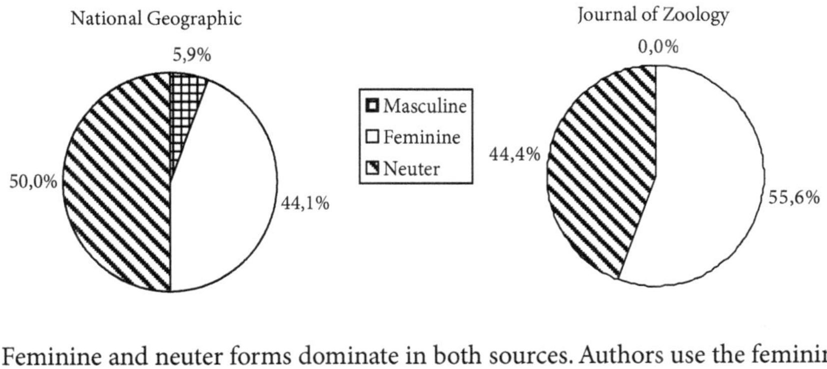

Feminine and neuter forms dominate in both sources. Authors use the feminine gender in describing reproduction and egg laying.

In the *National Geographic* corpus I have found four species with gender distinction. Masculine has been used only once (cf. C. 1.–4.)

Five snake species with gender assignment taken from *Journal of Zoology* papers show the dominant role of the female. All the examples retrieved refer to reproduction and egg laying. The terms as *mother, young, newborn* are used (cf. J. 18.–20.):

(109) [J.18.] **mother**'s body is 75% visible, resting on thermion probe, with none of **<u>her</u>** **young** visible.

(*Journal of Zoology* vol. 274, 2008, p. 279)

4.4.8 Bat

Common Vampire Bat, Big- Eared Bat, Epauleted Bat, North America's Bat, Panama's Bat, Free- tailed bat-*National Geographic*

Natterer's bat *(Myotis nattereri)*, European free-tailed bat *(Tadarida teniotis)* Short- nosed fruit bat *(Eidolon helvum)*, Daubenton's bat, Long-fingered bat *(Myotis capaccinii)*, Horseshoe bat *(Rhinolophus monoceros)*, Pipistrelle bat *(Pipistrellus pipistrellus)*, Mouse- eared bat *(Myotis myotis)*, Tent- making bat *(Artibeus watsoni)-Journal of Zoology*

Table 4.27: Comparative gender assignment – Bat

Pronouns	Masculine			Feminine			Neuter		
		NG	JZ		NG	JZ		NG	JZ
Personal subjective	he	1	1	she	1	6	it		20
Personal objective	him	4	1	her	4		it	16	3
Possessive	his	12		her	7	3	its	13	23
Reflexive	himself			herself	1		itself		5
TOTAL		17	2		13	9		29	51

<u>Material analysis</u>

Figure 4.29: Comparative charts.

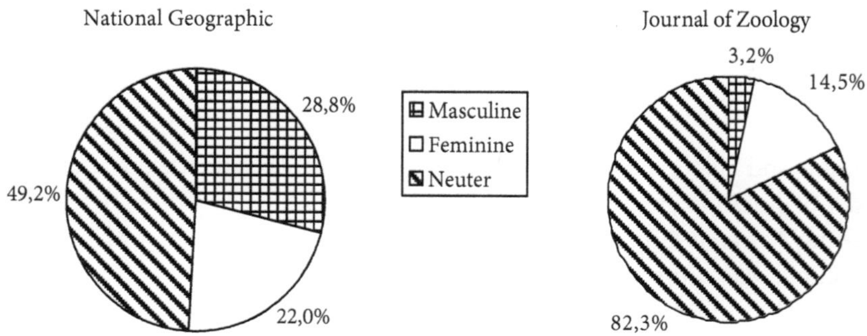

Six bat species taken from *National Geographic* have been assigned predominantly to the neuter. The use of masculine and feminine is nearly balanced. Masculine refers to observation and description of courting and mating, whereas feminine is used in the description of maternal care (cf. N. 15.–17.):

(110) [N. 17.] In June each female gives birth to a single pup. **She** leaves the cave twice each night to feed, returning to nurse the pup. .... To feed **herself** and **her** growing pup, a female bat must eat the equivalent of up to 70 percent of **her** body weight each night.

(*National Geographic* April 2002, p. 119)

In bats authors also pay attention to the young. They are referred to by ***it*** in all excerpted cases (cf. N. 18., 19.)

Bats belong to frequently -occurring species in research papers. I have discovered nine bat species with gender distinctions. Masculine is used in the description of parental care (cf. J. 21.), feminine and neuter in observation. The female under study is therefore assigned to ***she*** throughout the whole long-term study (cf. J. 22.):

(111) [J. 22.] The female was captured first as a juvenile and later recaptured 300m from **her** birth side..**She** was later recaptured on 3 March 2004 with a c. 2 month- old foetus, suggesting **she** was 6 months old when **she** mated.....

4.4.9 Albatross

Albatross- *National Geographic*

(Thalassarche cauta) -Journal of Zoology

Table 4.28: Comparative gender assignment – Albatross

Pronouns	**Masculine**			**Feminine**			**Neuter**		
		NG	**JZ**		**NG**	**JZ**		**NG**	**JZ**
Personal subjective	he		3	she			it		6
Personal objective	him		4	her		4	it		
Possessive	his		8	her	3	8	its	9	5
Reflexive	himself		1	herself			itself		1
TOTAL			**16**		**3**	**12**		**9**	**12**

Material analysis

Figure 4.30: Comparative charts – Albatross.

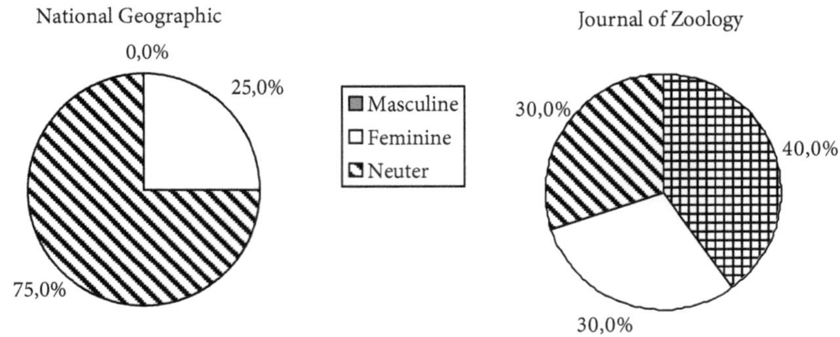

Although the majority of albatross species belong to threatened species, I have found only three articles in the corpus of *National Geographic* magazine concerning this animal. As can be seen from the above chart, neuter is dominant. The rest of the pronouns are feminine, referring to behaviour (cf.N.20):

(112) [N.20.] In the Falklands a female black-browed albatross yields **her** throat to **her** mate's nibbles, and offers a graceful beak-stroke of **her** own.
(*National Geographic* December 2007, p.101)

Gender assignment is quite different in research papers. The authors referred to males and females as *he* and *she* in the descriptions of behaviour, egg laying and nesting (cf.J.23.). In albatrosses both parents incubate the egg and feed the chick. Consequently, both feminine and masculine are used:

(113) [J.23.] females would climb onto the nest and sit if **she** was not already in this position. The male would then mount **her** and orient **himself** so that **he** was facing the same direction before lowering **his** body to rest against **her** back. The male would then lower **his** head and repeatedly tap **his** bill against the female's bill, while at the same time extending **his** tail out behind **him** and swinging it rapidly from side to side. During this time the female would deliberately raise **her** tail so that it came into contact with the male's tail, allowing **him** to eventually push **her** tail towards one side while bending **his** tail downwards to bring **his** cloaca into contact with **hers**. Following cloacal contact the female kept **her** tail raised and contracted **her** cloaca repeatedly. The male usually dismounted, although in rare instances **he** rested on top of the female for several minutes.
In one instance, a banded female that had been rejected by **her** paired male shortly before egg laying … Every case involved a breeding male attending **his** nest and a female that was not **his** paired mate. The male mounted or attempted

to mount either a breeding female attending a nearby nest or an unidentified bird moving past **his** nest….. the female kept **her** tail lowered in all cases …

(*Journal of Zoology* vol. 270, p. 630–631)

4.4.10 Seal

Elephant Seal, Fur Seal, Harp Seal, Hawaiian Monk Seal, Leopard Seal, Ringed Seal, Weddell Seal -*National Geographic*

Fur seal *(Arctocephalus forsteri),* Northern fur seal *(Callorhinus ursinus)- Journal of Zoology*

Table 4.29: Comparative gender assignment – Seal

Pronouns	Masculine			Feminine			Neuter		
		NG	JZ		NG	JZ		NG	JZ
Personal subjective	he	4		she	8		it	12	5
Personal objective	him			her	11		it		
Possessive	his	3		her	20	1	its	6	2
Reflexive	himself			herself			itself	3	
TOTAL		**7**			**39**	**1**		**21**	**7**

Material analysis:

Figure 4.31: Comparative charts – Seal.

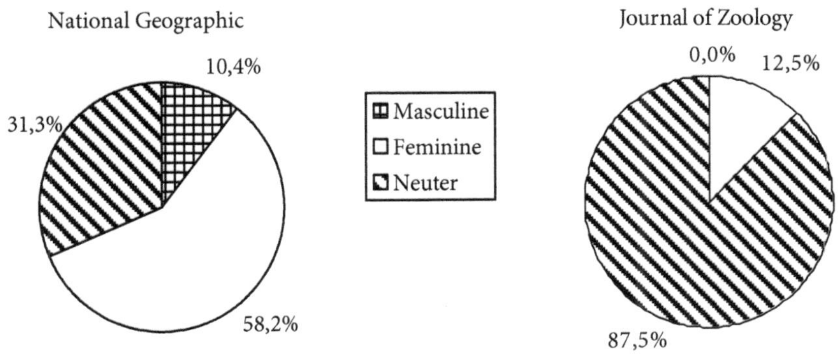

These semi-aquatic marine mammals are among frequently studied animals in both sources.

In *National Geographic* articles, the majority of examples refer to maternal care. I have found seven species. Most examples are devoted to the relationship of the mother and her pup. The authors pay special attention to the pup, because the seal gives birth to a single pup. Consequently feminine and neuter dominate (cf. N.33.–42.):

(114) [N.36.] **A mother monk** baahs to **her** black **pup**, which baahs back. For 40 days **she** will tend **her pup** without rating...

(*National Geographic* vol. 181, 1992, p. 141)

Masculine gender markers are used in descriptions of the male (cf. N. 30.,31.), and both masculine and feminine are used in examples referring to mating (cf. N.32.):

(115) [N.32.] When the male releases **his** grip, the female bolts, and **he** chases **her**...

(*National Geographic* vol. 181, 1992, p. 144)

Despite the fact that seals belong to frequently occurring species in *Journal of Zoology* research papers, I have found only two species with gender distinction. Neuter dominates in majority of retrieved examples. Only one female excluded from the research has been referred to as ***she*** (cf. J.25.):

(116) [J.25.] One 13-year-old female that was extremely small for **her** age and two males that were extremely large for their age were excluded from analysis......

(*Journal of Zoology* vol 272, 2007, p. 379)

4.4.11 Wild dog

African wild dog- *National Geographic*

African wild dog (*Lycaon pictus*)-*Journal of Zoology*

Table 4.30: Comparative gender assignment – Wild dog

Pronouns	Masculine			Feminine			Neuter		
		NG	JZ		NG	JZ		NG	JZ
Personal subjective	he	5	3	she	3		it	3	1
Personal objective	him	3		her	2		it		
Possessive	his	15	2	her	6	5	its	5	1
Reflexive	himself	1		herself			itself		
TOTAL		**24**	**5**		**11**	**6**		**8**	**2**

Material analysis

Figure 4.32: Comparative charts – Wild dog.

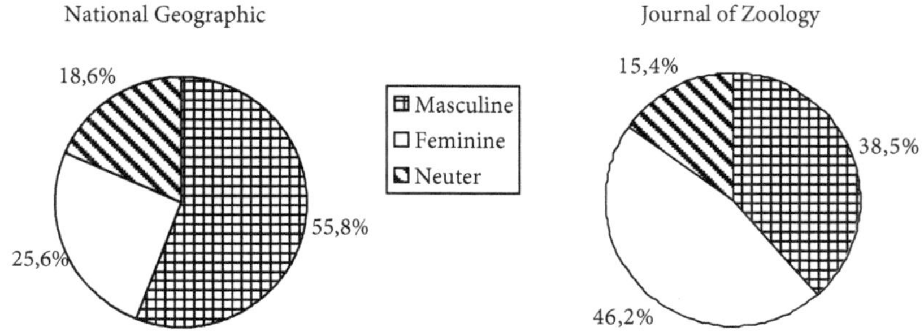

I have found this species of carnivorous mammals in both subcorpora. Although the charts above are not identical, the gender assignment is very similar, especially in the distribution of masculine markers. I have excerpted many examples referring to their life in packs with a specific social structure. Consequently, the authors use family expressions such as *mother, father, brother, sister, son, daughter*, when describing the dogs' hierarchy. All the individuals have names. The authors give names to the observed males and view them as ***he*** in both studied sources, as in the following pair of examples (cf. N.40.–42. and *Journal of Zoology* J.26.,27.):

(117) [N.41.] We find **Blackcomb** before the other dogs do, with **his** nose in the warm belly of an impala. **He** has made this kill by **himself**. **Blackcomb** leaves to bring in **his** pack mates. **He** suddenly stops and rears up on **his** hind legs. **His brother Tremblant** joins **him**.

(*National Geographic* May 1999, p. 53,54)

(118) [J.27.] It appeared that **Hans** did not accept **his** subordinate position, as **he** regularly refused to show submission to **George**.

(*Journal of Zoology* vol. 270,p. 506)

A similar situation has been observed in females. They have names and are viewed as ***she***. The examples refer to maternity care (cf. N.43.–46. and J.28.):

(119) [N.44.] **Bell**, the breeding female of **her** pack, is greeted by 11 pups.

(*National Geographic* May 1999, p. 39)

(120) [J.28.] The female and **her son** were released into the adjacent boma.

(*Journal of Zoology* vol. 270, p. 506)

As can be seen from the charts, neuter is marginal. The reason could be the hierarchy of the dogs' packs, where each individual is a unique member of the "family". Consequently, there is no need to use neuter.

4.4.12 Wolf

Arctic Wolf, Maned Wolf, Grey Wolf, Red Wolf, Indian Wolf, Wolf of Ethiopa, Wolverine-*National Geographic*

Maned wolf (*Chrysocyon brachyurus*), Grey wolf (*Canis lupus*), Indian wolf (*Canis lupus pallipes*)-*Journal of Zoology*

Table 4.31: Comparative gender assignment – Wolf

Pronouns	Masculine			Feminine			Neuter		
		NG	JZ		NG	JZ		NG	JZ
Personal subjective	he	18		she	15	4	it	7	2
Personal objective	him	11		her	10		it	1	
Possessive	his	19		her	15	16	its	14	3
Reflexive	himself		1	herself			itself	2	
TOTAL		48	1		40	20		24	5

<u>**Material analysis**</u>

Figure 4.33: Comparative charts – Wolf.

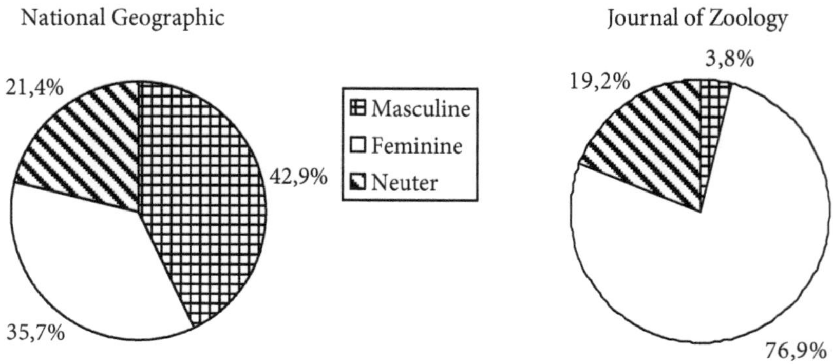

Although the wolf belongs to the Canidae family and has a very similar life cycle to the wild dog, the gender assignment is different.

In *National Geographic* articles both male and female gender assignment is used, predominantly in descriptions of parental care and social structure in packs. Although their family structure is not as unified as that of wild dogs, I have retrieved only smaller families with *mother, father and their offspring* (cf. N.47.–55.).

(121) [N.47.] Presently **the mother wolf** decided to take **her** pups for a walk.
.....I could only lie still, while the female hurried **her** pups back down the hill.
A wolf dashed over and trotted off with it, followed by **his** pack mates...
(*National Geographic* vol. 171, 1987, p. 570–571)

In research papers it is the feminine gender which dominates in three species. The majority of examples retrieved refer to observation of the studied female and her juvenile (cf. J.30.–34.):

(122) [J.32.] However, this asymptote may be an artefact because **the female** appeared to be searching for a nest site before giving birth and was therefore increasing **her** home range.. In the case of **the juvenile**, this may reflect the increasing physiological needs of a growing animal or **her** need to explore and gather information about **her** environment. For the 40 days before the birth of **her** pups, the female in our study was seen to be increasing and decreasing **her** home range. There are three possible explanations for this pattern of home range: (1) the female was procuring more food due to increasing energetic demands because of **her** pregnancy, (2) the female was procuring a birth site (nest/ den) and **her** increase in home range reflects search effort, and (3) the female was responding to decreased prey availability by increasing **her** home range.......
(p. 33)

The use of neuter in both sources observes similar rules as in wild dogs. The individuals are therefore viewed as ***he*** or ***she*** more often.

4.4.13 Fox

Arctic fox, Fennec fox, Red fox- *National Geographic*

Darwin's Fox *(Pseudalopex fulvipes)*, Hoary fox *(Pseudalopex vetulus)*, Red fox *(Vulpes vulpes)*, Island fox *(Urocyon sp.)*, Pampean fox *(Dusicyon gymnocercus) – Journal of Zoology*

Table 4.32: Comparative gender assignment – Fox

Pronouns	Masculine			Feminine			Neuter		
		NG	JZ		NG	JZ		NG	JZ
Personal subjective	he		3	she	1		it	3	2
Personal objective	him		1	her			it		
Possessive	his		5	her	1	2	its	2	9
Reflexive	himself		1	herself			itself		
TOTAL			**10**		**2**	**2**		**5**	**11**

Material analysis

Figure 4.34: Comparative charts – Fox.

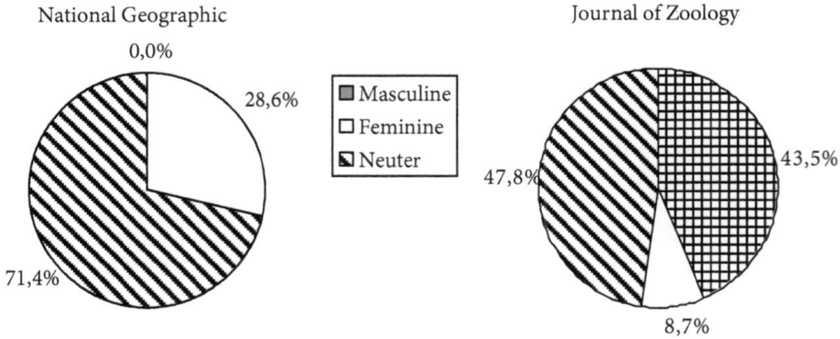

Three fox species are referred to predominantly by neuter. In the majority of examples, I have discovered general descriptions of life cycle. One situation where gender distinctions are used is maternity care. Authors thus speak about a mother and her young (cf. N.56.):

(123) [N.56.] When the pups were about two weeks old, their **mother** moved them one by one to a new den. Heavy rain may have prompted **her** decision, or **she** may have sought a cleaner site….

(*National Geographic* October 2004, p. 85)

Five species found in research papers are described mainly from the point of observation of their life cycle. The authors describe the family structure in more detail. They use both marks and names for the observed individuals (cf. J.35.–38.):

(124) [J.38.] **F5** was probably **F1**'s **daughter** from the previous season and may have helped **<u>her</u>** to raise their siblings.........**M2** was found wounded twice (**<u>he</u>** had an open wound on the shoulder and **<u>his</u>** left ear was dropped) in M0 range, whom **<u>he</u>** avoided......
M4 might have been paired with **F6**, but **<u>she</u>** wasn't breeding when caught. **M3** was likely not paired and **<u>he</u>** stayed on Penguin Island by **<u>himself</u>** for various periods of time.

(*Journal of Zoology* vol. 271, p. 69)

4.4.14 Lynx

Lynx- *National Geographic*

Bobcat *(Lynx rufus), Eurasian lynx (Lynx lynx)-Journal of Zoology*

Table 4.33: Comparative gender assignment – Lynx

Pronouns	Masculine			Feminine			Neuter		
		NG	JZ		NG	JZ		NG	JZ
Personal subjective	he	2	2	she	11	5	it	1	5
Personal objective	him			her	4	4	it		
Possessive	his	3	1	her	14		its		2
Reflexive	himself			herself			itself		
TOTAL		**2**	**3**		**29**	**9**		**1**	**7**

<u>**Material analysis**</u>

Figure 4.35: Comparative charts – Lynx.

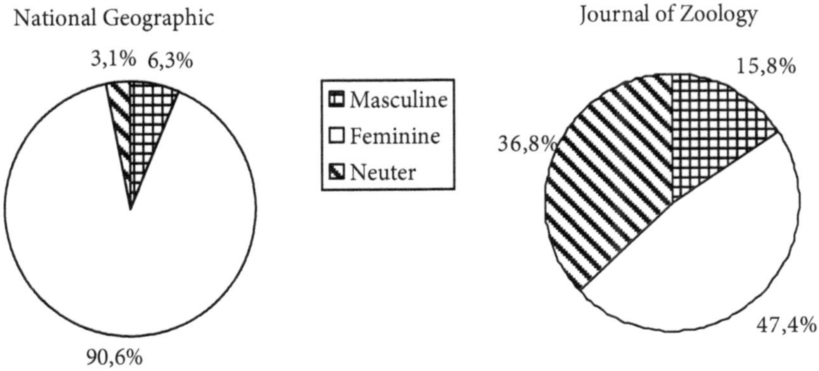

This cat is a solitary animal, thus descriptions of observation of only one individual can be found in both corpuses. I have also found examples referring to maternity care. The reason for the prevalence of feminine could be that the majority of retrieved animals are females.

In *National Geographic* articles authors name the monitored female, whereas in *Journal of Zoology* researchers give the male a number. In both examples below the animals are referred to by ***she*** and ***he***: (cf. N.60.–65. and J.39.–42.):

(125) [N.60.] As was **her** habit, the three-year-old female lynx padded solo through the deep snow. **She** spied a movement and pounced across a willow patch. **She** floated on **her** oversize paws through the snow. Dripping saliva, **she** chewed a branch coated with the fragrant paste. Suddenly something gripped **her** leg, jolting **her** out of **her** blissful state. Each time **she** moved, **she** dragged a cumbersome log, now wired to **her** left forefoot. **She** huddled warily, **her** tufted, pointy ears trained toward any sound that would reveal what awaited **her**. Other than that **she** was a healthy, 17-pound lynx. Before **her** odyssey was over, **she** would be examined. **She** would establish a new territory.

(*National Geographic* January 2006, p. 58)

Curiosity didn't kill **Chilkat**, as we'll call **her** after **her** capture location. But it sent **her** on a long journey.

(*National Geographic* January 2006, p. 59)

(126) [J.39.] **M53** was monitored during the first year and had a small home range with an 84% overlap with **his father**'s. However, the behaviour of this cat upon reaching breeding age was unknown because **he** died of unknown causes the first summer **he** was monitored…

(*Journal of Zoology* vol. 269,p. 499)

The occurrence of masculine is low in both materials. A possible explanation could be the prevalence of monitored female individuals. (cf. N.57.–59. and J.39.)

4.4.15 Bear

Black Bear, Brown Bear, Grizzly Bear, Great Bear, Polar Bear, Sloth Bear, Spectacled Bear, Sun Bear-*National Geographic*

Asiatic black bear *(Ursus thibetanus),* Andean bear *(Tremarctos ornatus),* Polar bear *(Ursus maritimus),* Brown bear (*Ursus arctos)-Journal of Zoology*

Table 4.34: Comparative gender assignment – Bear

Pronouns	Masculine			Feminine			Neuter		
		NG	JZ		NG	JZ		NG	JZ
Personal subjective	he	12	1	she	20	4	it	6	8
Personal objective	him	5		her	5		it	1	
Possessive	his	11	2	her	21	2	its	13	2
Reflexive	himself			herself	2		itself		
TOTAL		**28**	**3**		**47**	**6**		**14**	**10**

Material analysis

Figure 4.36: Comparative charts – Bear.

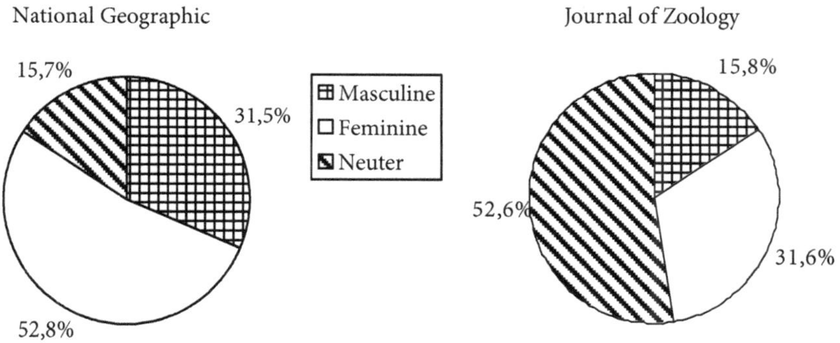

In this analysis I study both brown and polar bear species- eight of them excerpted from *National Geographic*, four from *Journal of Zoology*.

In both materials the majority of examples retrieved refer to maternity care and the relationship between the mother and her cub (cf. F. 1.–16. and J. 44., 45.). The female is assigned to the feminine gender as ***she***:

(127) [F. 3.] A lone male warily eyes a female guarding **her** young. **She** emerges with **her** cubs in spring, when **she** can hunt newborn seals to replenish **her** fat reserves.

(*National Geographic* January 1998, p. 70)

(128) [J. 44.] We suspect that this was related to **her** occupying a birthing den that blocked radio transmission. In November 2000 we observed this bear and, although we did not see cubs, **she** made a soft cooing sound, which suggested **she** was calling **her** cubs.

(*Journal of Zoology* vol. 271, p. 205)

Masculine gender markers appear in descriptions of the life cycle as well as in the observation of solitary male bears (cf. F. 17.–23. and J. 43.).

4.4.16 Horse

Miniature horse, Mustang-*National Geographic*

Przewalski horse *(Equus ferus przewalskii)-Journal of Zoology*

Table 4.35: Comparative gender assignment – Horse

Pronouns	Masculine			Feminine			Neuter		
		NG	JZ		NG	JZ		NG	JZ
Personal subjective	he	5	2	she	4		it	6	2
Personal objective	him			her		1	it		
Possessive	his	14		her	16		its	4	3
Reflexive	himself			herself			itself		
TOTAL		**19**	**2**		**20**	**1**		**10**	**5**

Material analysis

Figure 4.37: Comparative charts – Horse.

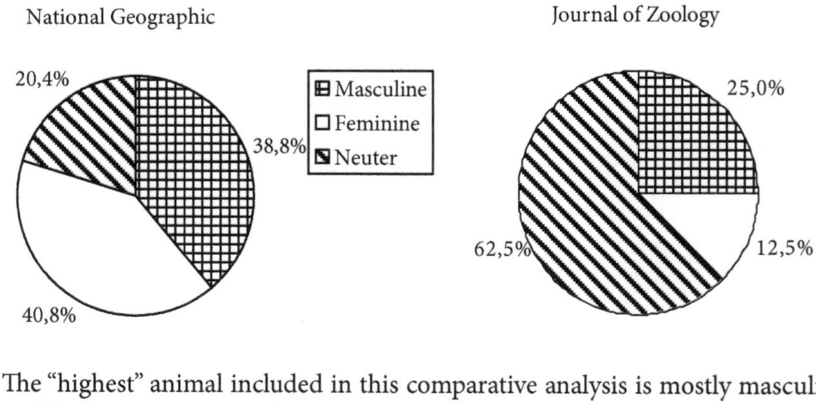

The "highest" animal included in this comparative analysis is mostly masculine and feminine in *National Geographic* corpus and neuter in *Journal of Zoology* papers.

The examples mostly refer to the behaviour and breeds of horses. I have retrieved two breeds of wild horses – Mustang, retrieved from *National Geographic*, and Przewalski horse in *Journal of Zoology*. Both are wild and endangered and thus predominantly assigned to the masculine and feminine in both sources. They have names. The majority of horses are stallions, as can be seen in two following examples, taken from both corpora (cf. N.66.–70. and J.45.):

(129) [N.68.] **The mustang** was white all over, with dark spots on **his** rear, betraying some Appaloosa in **his** ancestry. **He**'d been named **Dot** by the inmates. **Dot** lowered **his** head to smell for prairie dog and badger holes to keep from falling and, compensating with brains and courage for what **he** lacked in beauty, took the terrified youngster right back to camp.

(*National Geographic* February 2009, p. 105)

(130) [J.45.] If the **stallion** is in good physical condition **he** will be less likely to be confronted and lose the mare, and the mare will not be harassed by a large number of stallions attempting to mate with **her**.
Stud pile marking throughout the home range advertises site-specific dominance by the stallion in a cost-effective manner and the fact that **he** is the resource holder.

(*Journal of Zoology* vol. 271, p. 3)

4.4.17 Crayfish, Possum, Mongoose

The three species mentioned above have gender distinction only in research papers. I have not found any pronoun referring to gender in *National Geographic* articles thus they have not been included into this analysis. The appropriate tables and examples can be found in the appendix.

Freshwater crustacean **Crayfish** together with mantis belong to lowest animals with gender distinction in research papers. Crayfish is assigned to the feminine by ***she*** in all excerpted examples referring to the reproductive process (cf. J.46.,47.):

(131) [J.46.] Therefore, the female choice of larger males might reflex the advantages that the female **herself** gains from mating with them.....

(*Journal of Zoology* vol. 274, 2008, p. 176)

Arboreal marsupial **Possum** is viewed as ***he*** and ***she*** in the long-term research described in research papers (cf. J.48.,49.):

(132) [J.48.] We considered a **male** to be adult. Once **his** scrotal width was 35 mm and **he** was assigned as >TWC 2. We considered a **female** to be adult once **she** had bred for the first time..... We considered animals to be juveniles until the age of I year the approximate time at which young were first trapped separately from their mothers.

(*Journal of Zoology* vol. 271, 2007, p. 378)

The last animal with gender markers found only in *Journal of Zoology* corpus is the mammal **Mongoose**. The observed male and female are viewed as ***he*** and ***she*** (cf. J.50.,51.):

(133) [J.50.] The only **dominant female** that we followed had a home-range size of 0.20 km near the centre of **her** mate's territory.....
However, the area where the dominant female deposited most of **her** cheek marks was a location where **her** mate concentrated a high number of cheek marks. **Her** anal marks were in an area where the male also marked anally, but the closest male anal mark was 20m from **her** marks...

(Journal of Zoology vol 275, 2008, p. 35)

4.4.18 Final charts and material analysis

16 species taken from *Journal of Zoology* were referred to by 413 pronouns as follows:

Masculine: 57 = 13.8%
Feminine: 168 = 40.7%
Neuter: 188 = 45.5%

Total: 413 = 100%

16 species taken from *National Geographic* were referred to by 752 pronouns as follows:

Masculine: 176 = 23.4%
Feminine: 287 = 38.2%
Neuter: 289 = 38.4%

Total: 752 = 100%

Figure 4.38: Final comparative charts.

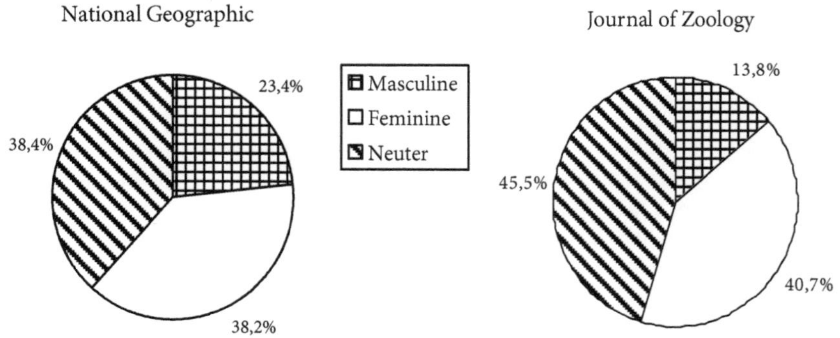

The above comparative analysis includes a wide variety of animal species. All of them contain a surprisingly high number of different gender markers. Although the number of the pronouns is higher in *National Geographic* articles, the percentage of occurrence of all masculine, feminine and neuter gender markers in both sources is, surprisingly, nearly in balance. The most illustrative example is the comparison of the "lowest" animals of this comparative analysis – the insect species, especially the **praying mantis.** The charts show the same proportions of masculine pronouns and a very similar gender occurrence of feminine and neuter markers. The occurrence of all gender markers was nearly the same in both sources. The main themes of all articles concerning insects were courtship and mating processes. Above all, the reproductive process of the praying mantis is so specific that all the excerpted articles had nearly the same structure and gender references.

Different proportions of gender assignment have been found in another researched species of insets- **cricket**. Although the number of species in both sources was the same, the occurrence of gender markers were noticeably higher in research papers.

Parental care was mostly mentioned in reference to **spiders**. The authors used terms such as *mother, spiderlings, young*, etc., that would not be expected in scientific papers.

The **turtle species** was a typical example of an animal in a long-term research study which is always assigned to gender. The prevalence of feminine pronouns in *Journal of Zoology* corpus was caused by the long-term study of the analysed female.

The **reptiles** also show balanced gender assignment in both sources. It could be explained by the same themes under study, concerning mating and motherhood.

Bat species belong to the most frequently occurring animals in both sources. I have discovered 20 species of this animal in the research papers, nine of them with gender references mostly related to parental care. The majority of the pronouns in both corpora were neuter.

The next group under study was carnivorous mammals. I have found gender distinctions in four species- **wild dog, wolf, fox** and **lynx.** The high occurrence of masculine and feminine pronouns could be explained by their life in packs with a specific social structure and hierarchy. The researchers have even named the animals. Thus life in the "family" influences gender markings in all the examples analysed. In the analyses of solitary mammals such as **lynx** and **bear**, the majority of examples referred to the relationship between the female and her young. Consequently the feminine gender prevailed.

The **horse** was the "highest" animal included into this comparative analysis. I have retrieved only five articles concerning horse species. The gender assignment in horses has been connected with description of breeds and behaviour. The majority of them have names. Stallions have prevailed.

Three species of animals found in research papers have no gender distinctions in *National Geographic* articles: freshwater crustacean **Crayfish**, arboreal marsupial **Possum** and mammal **Mongoose**.

To sum up, all the species show a surprisingly high number of gender markers in both of the sources analysed, especially in the zoological research papers. Females with their young have been mostly mentioned in the research papers. They are probably more important for scientific purposes.

The main contexts and situations where observing strict gender distinctions proved important were similar in both sources:

- The description of mating and courtship processes.
- The description of motherhood and parental care.
- The close relationship between the researchers and the animal in long-term studies.
- The animals analysed are identified through markings.

Although the biological sex plays the dominant role in assigning gender to animals, the sociological aspect in both study materials is also important.

4.5 The analysis of gender in juvenile animals

The aim of this chapter is to analyse and compare gender assignment in juvenile animals retrieved from *National Geographic* articles and scientific papers in *Journal of Zoology*. It attempts to ascertain differences -if any- of gender assignment in youngsters within the two subcorpora. The initial hypothesis was that gender assignment to juvenile animals would manifest differences from that in adults.

The two tables below list all the young discovered with their gender references. I have collected two sets of materials:

- The first subcorpus represents 30 volumes of the English edition of *National Geographic* (1979–2009).
- The second subcorpus includes three volumes of the English edition of *Journal of Zoology* (2006–2008).

The two tables below list all the young discovered along with their gender references. The expressions referring to each juvenile animal are given in brackets.

4.5.1 Youngsters in National Geographic

Table 4.36: Gender assignment in youngsters – National Geographic

Youngster	Masculine	Feminine	Neuter
Antelope (calf)			2
Aphid (newborn)			2
Baboon (baby, infant)			3
Bat (pup, young)			4
Batfish (juvenile)			2
Bear (cub, young)			1
Birds (juvenile, chick)	8	1	14
Bonobo (firstborn)	3		
Cuttlefish (newborn, young)			2
Eagle (juvenile, youngster, eaglet, chick)	46		4
Earwig (young, offspring)			1
Elephant (calf, juvenile, baby, offspring)	9	6	4
Frog (froglet, tadpole)			3
Giraffe (baby)			1
Gorilla (offspring, infant, youngster, baby)	22	8	5
Honey Badger (cub)	5	1	1
Hyena (young, cub)			2
Jaguar (young, cub)	16		
Leopard (cub, little one)		37	
Lion (cub, kitten)	7		2
Lionfish (juvenile)			3
Lizard (youngster)		8	
Lynx (kitten, newborn)		2	
Mola mola (baby)			2
Octopus (young)			2
Owl (owlet, chick)			1

Youngster	Masculine	Feminine	Neuter
Panda (cub, young, newborn)	11		3
Pearlfish (young, nestling)			1
Seal (pup, baby, youngster)		2	1
Shark (pup, newborn, young, baby, yearling)			7
Snake (youngster)		4	
Tiger (cub, young)	11	2	
Whale (calf, juvenile, baby, yearling)	8	3	30
Wolf (pup, young)	12		1
TOTAL	**158**	**74**	**102**

4.5.2 Youngsters in Journal of Zoology

Table 4.37: Gender assignment in youngsters – Journal of Zoology

Juvenile animals	Masculine	Feminine	Neuter
Albatross (chick)			1
Baboon (infant)			1
Badger (cub)			1
Bat (young, pup)			1
Bear (cub, yearling)		2	1
Beaver (kit)			1
Bobcat (juvenile, offspring)		2	
Bobuck (young)	1	1	
Civet (young)			2
Cougar (Cub)			1
Crawfish (juvenile)			1
Deer (Fawn)			1
Elephant (calf, juvenile)			2
Flamingo (young)			1
Fox (pup, offspring)			1
Frog (juvenile)			1
Gazelle (fawn)			2
Gecko (neonate)			1

Juvenile animals	**Masculine**	**Feminine**	**Neuter**
Ground squirrel (juvenile, young)			1
Grouse (Chick)			2
Guinea-pig (pup)			1
Gull (Chick)			1
Hedgehog (young)			1
Ibex (kid)			1
Impala (young)			1
Kangaroo (young)			2
Koala (Joey)			2
Lizard (hatchling)			1
Lynx (kitten, cub)			1
Mongoose (pup)			10
Monkey (juvenile, infant)			1
Otter (juvenile)			2
Penguin (Chick)			1
Possum (young, juvenile)	2		
Rattlesnake (young)			1
Roe deer (fawn)			1
Salamander (offspring)			1
Salmon (young, juvenile, offspring)			2
Seal (pup)			2
Snake (hatchling, juvenile)			1
Spider (spiderling)			1
Turtle (nestling, hatchling)			2
Wild dog (pup)			2
Wolf (juvenile, pup)		3	5
Wolverine (kit)		2	2
TOTAL	**3**	**10**	**67**

4.5.3 Material analysis of gender in juvenile animals

33 youngster species taken from *National Geographic* were referred to by 334 pronouns as follows:

Masculine:	149 = 47.3%
Feminine:	81 = 22.4%
Neuter:	84 = 30.5%
Total:	334 = 100%

45 youngster species taken from *Journal of Zoology* were referred to by 80 pronouns as follows:

Masculine:	3 = 3.8%
Feminine:	10 = 12.5%
Neuter:	67 = 83.8%
Total:	80 = 100%

Figure 4.39: Comparative charts – Juvenile animals.

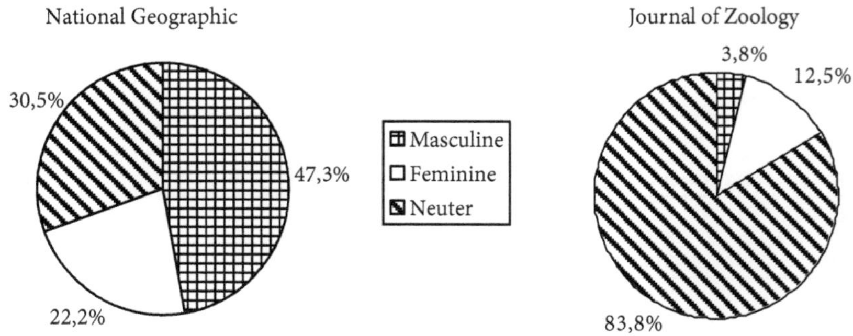

The results, as can be seen from the above charts, show the tendency to prefer masculine in National Geographic articles and neuter in *Journal of Zoology* corpus when referring to the young.

The preference for masculine in the case of the young taken from *National Geographic* corpus is evident. The dominance of masculine markers can be explained by the high occurrence of masculine markers especially in four species (eagle, gorilla, jaguar and panda). All these species belong to threatened animals and therefore get a lot of attention. For example, the **eaglet** is always referred to by *he* in the *National Geographic* corpus. He was observed for a year and researchers gave him a name (cf. Y. 17.–23.):

(134) [Y.21] "On May 6,1978, when **he** was 4,5 months old, **Doofus, the juvenile**, left **his** nest for the last time. The biologist saw **him** resting 700 yards from **his** natal tree. There **he** stayed, fed by **his** parents, until late August, when **he** strayed toward the border. The parents kept locating the youngster and bringing **him** food. One day I saw **young Doofus** fly to a knothole in a tree. Comically, **he** grasped the rim with **his** feet, using **his** tail as a prop. Then, wrapping **his** wings around the trunk, **he** poked **his** head into the cavity. Suddenly, **his** head recoiled, and one of **his** long legs shot down into the hole. **He** squeezed and bit the chunk.

(*National Geographic* July 1981 p. 853)

A similar gender assignment can be seen in the **panda** species. Panda is among the world's most adored and protected rare animals with a very low birth rate (*National Geographic* July 2006). That is why researchers personalise panda cubs, giving them names and talk about them as ***he*** and ***she*** (cf. Y. 62.–66.):

(135) [Y.62.] "**He**'s got chubby cheeks. **He** naps a lot. **He** eats with **his** hands. **He** lives with **his** mother. **Tai Shan** is **a giant panda cub**, and that makes **him**, well, not your average bear. Born at 3: 41 a.m. on Saturday, July 9, 2005, at the Smithsonian National Zoological Park in Washington, D.C. **Tai Shan** is the first **offspring** of Tian Tian. Internet panda cams, which follow the daily activities of **Tai Shan** and **his** mom, draw an average of two million online visits a month.

(*National Geographic* July 2006, p. 48)

The same gender distinction can be seen in young **gorillas**. They have names or numbers and are viewed as ***he*** or ***she*** (cf. Y. 31.–45.):

(136) [Y.41.] "When we first met Beth in 1971, she was carrying **infant Bert** on her neck. Bert must have been approximately a year old, because **he** was briefly leaving **his** mother's body for short distances.

(*National Geographic* June 1980, p. 837)

The next group of youngsters with a high occurrence of feminine and masculine gender markers are "**big cats**". I have discovered only two examples of neuter references in the description of maternal care (cf. Y. 57., 58.):

(137) [Y.57.] "Ever so carefully, a lioness slides her enormous canine teeth around her **five-week-old cub**'s head before picking up **the youngster.** She carries **it** 15 feet, then puts **it** down and moves on ahead, calling all the while with soft, moaning grunts. Finally **the cub** responds to her coaxing and trots after her.

(*National Geographic* December 1982, p. 811)

The rest of the cubs are of masculine or feminine gender.

All **jaguar cubs** are viewed as ***he*** (cf. Y.48.):

(138) [Y.48.] "**A young male jaguar** rises from **his** sleep, stretches, and silently but determinedly leaves forever the place where **he** was born.
(*National Geographic* March 2009, p. 123)

The **leopard cub** Legadema, observed for 13 months, is viewed as ***she*** (cf. Y.49.–56.):

(139) [Y.50.] "**She** was eight days old when we spotted **her**. **Her** eyes were still milky gray, and **she** wobbed slightly. Emerging into the sunlight from **her** den, **she** seemed curious and bold. **Her** mother had lost five previous cubs to hyenas, baboons, and other predators. What would happen to this one? We decided to follow **the little one** as **she** grew up. From **her** first days, Legadema, as we came to call **her**, was under constant threat. Whenever Legadema went, vervet monkeys with darting eyes spotted **her** a mile off. **Her** mother, a patient teacher, instructed Legadema in the skills **she** would need to survive as a predator.
(*National Geographic* April 2007, p. 144)

The same gender assignment is visible in **tiger cubs**. Both male and female youngsters have names and are used only with masculine and feminine gender markers (cf. Y.72.–76.):

(140) [Y.73.] "Perched outside the den where **he** was born, **a male cub** named **Globus** surveys the enclosure. **He** and a brother were born to Koucher and Niurka. Despite great efforts to save Globu's brother, **he** died of malnutrition. Though malnourished **himself**, Globus survived, thanks to the care of Victor Yudin. In October 1996 we took Globus to the Minnesota Zoo. At maturity **he** will add a welcome infusion of new genetic material to the captive-management program. Since Globus's birth, **his** parents have produced three more cubs.
(*National Geographic* February 1997, p. 109)

I have discovered many masculine and feminine gender markers in other species of mammals such as **whale calves** (cf. Y.77.–92.), **seal pups** (cf. Y.68.), **elephant calves** (cf. Y. 24.), **wolf pups** (cf. Y. 93.), **lynx kittens** (cf. Y.60.) and **honey badger cubs** (cf. Y.46.).

The "lowest" animals with masculine or feminine gender assignment in their youngsters are **lizard** and **rattlesnake** species. The following quotations refer to observation of these animals. The authors use more personal and emotive feminine when describing lizard and rattlesnake youngsters (cf. Y.59., 67.):

(141) [Y.59.] "Well, **she** seems to have had **her** Wheat-a-Bix this morning," Dickman says. "**She** is just **a youngster**; no pouch development yet," he adds, showing me the soft pink circle on **her** stomach that will one day hold **her** young. He weighs and measures **her**, then carefully clips tiny triangles of skin from the edge of **her** ears to mark **her**.
(*National Geographic* February 1999, p. 27)

(142) [Y.67.] I had marked this female timber rattlesnake as **a youngster**. **She** was the first I'd ever seen in the wild. I remembered **her** number. I had wondered if I would ever see **her** again.
Warily I slipped the venomous reptile into a nylon bag, attached an identifying tag and put **her** in my backpack with others to be carried down to the mountains.
(*National Geographic* July1987, p. 30)

Eleven species of youngsters are **neuter**. The lowest animals with neuter reference in youngsters are
aphids (small plant-eating insects) (cf.Y.2.):

(143) [Y.2.] "This samurai aphid species also lacks soldiers, yet any **newborn** can use **its** horns to crush a predator's eggs.
(*National Geographic* September 1989, p. 412)

frogs (cf.Y.28.,29.):

(144) [Y.28] "Life and strife begin together as **a tiny froglet** hatches fully formed, bypassing the tadpole stage. I watched one for two days straight, but **it** ate nothing.
(*National Geographic* March 2004, p.27, *National Geographic* May 2001p.107)

(145) [Y.29.] The Phasmahyla guttata **tadpole** wears **its** mouth like a hat, allowing **it** to skim food from the surface of mountain streams.
(*National Geographic* March 2004, p. 27)

octopus (cf.Y.61):

(146) [Y.61.] "They find protection against stealthy predators, such as **a young five-inch-long octopus**, which alters **its** colouring by contrasting or expanding chromathophores- pigment sacs. Light reflects off **its** internal organs, shielded by a bulbous transparent mantle.
(*National Geographic* December 1981, p. 836)

It is surprising, that such "low" organisms as aphids (small plant-eating insects) and tadpoles (the larval stage in the life cycle of an amphibian) appear in singular and are therefore viewed as ***it***.

The majority of young animal names in the papers of the *Journal of Zoology* appeared in plural. Youngsters in singular occurred very rarely. I have recorded 45 species of youngsters which appeared in singular.

When referring to young animals taken from the *Journal of Zoology*, the authors assigned gender to all 45 retrieved species of youngsters. I divided them into four groups according to gender references:

Animals used only with masculine reference:

Possum (cf.Y.111.,112):

(147) [Y.112] "Once **his** scrotal width was 35 mm and **he** was assigned as >TWC 2. We considered animals to be juveniles until the age of I year the approximate time at which young were first trapped separately from their mothers.

(*Journal of Zoology* vol. 271, 2007, p. 378)

Animals used only with feminine reference:

Bear (cf. Y. 95., 96.):

(148) [Y. 96.] "As **a yearling she** was active 66% during the day, whereas as subadult she was active 46% during the day...

(*Journal of Zoology* vol. 269, p. 479)

Bobcat (cf. Y. 98., 99.):

(149) [Y. 98.] "Females established home ranges at WWR and subsequently each bred and mothered at least **one offspring. Juvenile** F48 remained on WWR through **her** second year, but died that summer...

(*Journal of Zoology* vol. 269, p. 500)

Animals with feminine and neuter reference:

Wolf:

(150) [Y. 122.] "**The juvenile female** visited **her mother** on several conditions... The relationship between **the juvenile** and **her parents** is intriguing and may suggest that **the juvenile** had an important role in the reproduction success of **her parents**... This was obviously in response to **her** need to dedicate time for parental care of **her pup**.

(*Journal of Zoology* vol. 271, p. 34)

Wolverine (cf. 123., 124.):

(151) [Y. 124.] "The relationship between **the juvenile** and **her** parents is intriguing and may suggest that **the juvenile** had an important role in reproductive success of **her** parents.... **the juvenile female** visited her mother......

(*Journal of Zoology* vol. 271, p. 34,35)

Animals used only with neuter reference:

This group comprises the majority of animals (see the table above) as in the case of **baboon** (cf. Y. 93., 94., 105.):

(152) [Y. 94.] "**Infant 4** essentially never utilized diagonal sequence gait, but **its** shift is consistent....no **infant** shows a preference for lateral couplets."

(*Journal of Zoology* vol. 269, p. 197)

The authors use expressions as **offspring, young, juvenile, sibling, newborn, neonate, or** they mark the researched youngsters when referring to them in the papers. (cf. Y. 94.)

The "lowest" animal retrieved from the *Journal of Zoology* corpus is the **spider**. The researchers used a variety of expressions as *young, offspring, juvenile and spiderling* when referring to young spiders (cf. 119.–122.):

(153) [Y.120.] "…together with as little variation as 9–116 cm from **one spiderling** to another from the same natal nest."

(*Journal of Zoology* vol. 268, p. 124)

The following two quotations concerning observations of young bears illustrate similar usage of gender markers in both corpuses (cf. Y. 96., 97. and Y. 5., 62.–66.):

(154) [Y.66.] **A young Wolong male** munches bamboo provided by keepers. **He** may soon join other captives being trained to be more self-sufficient.

(*National Geographic* July 2006, p. 54)

(155) [Y.96.] As **a yearling she** was active 66% during the day, whereas as subadult she was active 46% during the day…

(*Journal of Zoology* vol. 269, p. 479)

To sum up, even if there is a strong disproportion between the number of gender markers taken from both sub corpora (334 markers taken from *National Geographic* and 80 markers taken from *Journal of Zoology*, on the other hand the number of youngster species was surprisingly higher in *Journal of Zoology* corpus- 45 species whereas *National Geographic* yielded only 33 species of youngsters with gender references), the general rules of gender assignment in juveniles are very similar.

In spite of the fact that the two sub corpora are very different, the principles of gender markings are similar.

From the above data, the gender distinctions in youngsters were used predominantly in the descriptions of

- **parental care** and **family relationships**
- **research** and **observation** connected with the particular youngster.

While grammatical gender and natural sex play a very important role in assigning gender to youngsters, the **emotional involvement** of the researchers is apparent in the majority of examples.

4.6 The Comparative analysis of the use of gender markers by Czech students and native speakers

The corpus for this analysis consists of responses by 127 Czech students and 12 responses by native speakers.

The aim of this analysis was to ascertain the differences in gender assignment system in English and in Czech. As the first step I have compared the original articles concerning animals written by English authors and their Czech translations. I have chosen articles containing 21 species, which were included in 14 volumes of both English and Czech editions of *National Geographic* (2002–2015).

I have selected 21 various species of lower and higher animals, comprising invertebrates, insects, birds and mammals *(acoel, caterpillar, worm, jellyfish, mantis, salmon, scorpion, albatross, crane, wolf, honey badger, kinkajou, lynx, hyena, leopard, moose, snow fox, bear, panda, orca and chimpanzee)*.

I have excerpted 21 short texts referring to each species both in English and Czech:

<u>Acoel (Ploštěnka)</u>

An acoel, a tiny flatworm without a gut, inhabits stony shallows. Photosynthetic algae in **<u>its</u>** skin help nourish **<u>it</u>**.
National Geographic February 2007, p. 123

Drobná ploštěnka obývá kamenité mělčiny. S výživou jí pomáhají fotosyntetizující řasy, které se nacházejí v její pokožce.
National Geographic Únor 2007, str. 41

<u>Caterpillar (Housenka)</u>

In one-twelfth of a second the caterpillar whips back and ensnares the termite, then gobbles **its** meal. Belly full, **it** hangs languidly and finishes off the last bits while antennae on the termite's severed head twitch nearby.
National Geographic June 2003, p. 103

V jedné dvanáctině sekundy housenka termita polapí, pak metodicky pozře. S plným břichem se malátně zavěsí, zatímco tykadla na utržené hlavě termita poblíž sebou stále škubou.
National Geographic Červen 2003, str. 101

<u>Worm (Rournatec)</u>

This reef dweller protects **<u>its</u>** six-inch body in a sand-and-mucus tube while **<u>its</u>** appendages drift, snagging detritus and distracting foes.
National Geographic February 2007, p. 118

Chrání své 15 cm dlouhé tělo v trubičce z písku a slizu, zatímco jeho, vlající chapadélka zachytávají částečky odumřelé organické hmoty a odhánějí nepřátele.
National Geographic Únor 2007, str. 38

<u>Jellyfish (Medůza)</u>

This jellyfish was dubbed „big ugly" when **<u>it</u>** was discovered. **<u>It</u>** can grow to more than three feet.
National Geographic June 2004, p. 48

Tato medůza dostala přezdívku „velká ošklivá", když byla objevena. Může dorůst do více než metrového průměru.
National Geographic Červen 2004, str. 98

Mantid (Kudlanka)
She seems almost human, this mantid I found in West Africa. **She** has such alert eyes, and **her** head tilts to follow me.
National Geographic January 2006, p. 103

Tato kudlanka, kterou jsem našel v západní Africe, vypadá téměř lidsky. Má takové živé oči a naklání hlavu ke straně, aby mě mohla sledovat.
National Geographic Leden 2006, str. 49

Salmon (Losos)
After defending **her** nest for as long as **she** can swim upright, a salmon drifts back into an eddy to die.
National Geographic February 2003, p. 124

Losos brání své hnízdo tak dlouho, dokud dokáže plavat ve vzpřímené poloze, pak je stržen vírem a umírá.
National Geographic Únor 2003, str. 96

Scorpion (Škorpion)
In some, like this Arizona species , the scorpion stands guard over **her** eggs and later **her** newborns.
National Geographic July 2004, p. 99

U některých druhů, například u tohoto z Arizony, matka hlídá vajíčka a později také vylíhnutá mláďata.
National Geographic Červenec 2004, str. 71

Albatross (Albatros)
When the parent arrives, **it** and the chick excitedly position their bills crosswise. An adult may spend 15 minutes ashore feeding **its** youngster.
National Geographic December 2007, p. 107

Jakmile rodič dorazí k hnízdu, mládě nadšeně natočí zobák tak, aby se zkřížil s rodičovským zobákem. Stráví na břehu až patnáct minut a po celou dobu krmí mládě.
National Geographic Prosinec 2007, str. 125

Crane (Jeřáb)
On Hokkaido, Japan, a yearling red-crowned crane flares **its** wings and bows **its** head in a dance of either courtship or aggression.
National Geographic April 2004, p. 39

Jednoletý jeřáb mandžuský na japonském ostrově Hokkaidó rozpíná křídla a pokyvuje hlavou při tanci, jímž vyjadřuje dvoření nebo agresi.
National Geographic Duben 2004, str. 65

Wolf of Ethiopia (Vlček Etiopský)
Remember that young wolf who's babysitting the pups at the den. **She**'s at the age where it's time for **her** to move out and find **her** own pack.
National Geographic March 2006, p. 131

Vzpomínáte si na tu mladou samici, která hlídá štěňata v doupěti? Je ve věku, kdy by se měla osamostatnit a najít si vlastní smečku.
National Geographic Březen 2006, str. 63

Kinkajou (Kinkažů)
Kinks usually bear one pup every year or two, nursing **it** for as long as four months.
National Geographic October 2003, p. 53

Kynkažuové obvykle rodí jedno mládě ročně a kojí je až čtyři měsíce.
National Geographic Říjen 2003, str. 83

Honey badger (Medojed)
One night we saw a young badger collapse. **He**'d been struck in the face by a puff adder just before **he** bit its head off.
National Geographic September 2004, p. 103

Jedné noci jsme viděli, jak se zhroutil mladý samec, kterého do tváře zasáhla zmije útočná těsně před tím, než jí ukousnul hlavu.
National Geographic Září 2004, str. 113

Lynx (Rys)
A lynx in Canada, **his** right forefoot caught in a snare, awaits **his** fate.
National Geographic February 2006, p. 69

Rys s pravou přední tlapou zachycenou v oku čeká na svůj osud.
National Geographic Leden 2006, str. 69

Hyena
This young hyena has no intention of sharing. **It** could lose **its** lunch entirely.
National Geographic June 2005, p. 59

Mladá hyena nemá vůbec v úmyslu se dělit. O svůj oběd by nicméně mohla přijít úplně.
National Geographic Červen 2005, str. 43

Leopard
When the cub was five months old, **her** mother brought **her** a live baby impala.
National Geographic April 2007, p. 148

Když bylo mláděti pět měsíců, přinesla jí matka živé mládě impaly.
National Geographic Duben 2007, str. 136

Moose (Los)
He no longer shook **his** antlers at his assailants.
National Geographic May 2004, p. 104

Los se už přestal ohánět parohy po útočnících.
National Geographic Květen 2004, str. 50

Snow Fox (Polární liška)
Heavy rain may have prompted **her** decision, or **she** may have sought a cleaner site.
National Geographic October 2004, p. 85

K jejímu rozhodnutí možná přispěl vydatný déšť nebo možná hledala čistější místo.
National Geographic Prosinec 2004, str. 101

Bear (Medvěd)
A week or two out of hibernation, **he**'s spending the day filling **his** belly on the first lush green of spring.
National Geographic February 2006, p. 50

Před jedním či dvěma týdny se probudil ze zimního spánku a svůj den strávil tím, že si plnil žaludek první jarní šťavnatou zelení.
National Geographic Únor 2006, str.108

Panda
Tai-Shan will stay at the National Zoo with **his** mother until **he** is two years old.
National Geographic July 2006, p. 49

Tai Shan zůstane ve Washingtonu s matkou nejméně do svých dvou let.
National Geographic Červenec 2006, str. 81

Orca (Kosatka)
Rather, **she** is herding the fish until **her** calf joins the hunt.
National Geographic April 2005, p. 104

Kosatka se nijak zvlášť nesnaží rybu ulovit, spíše ji honí a čeká, až se k lovu připojí její mládě.
National Geographic Duben 2005, str. 126

Chimpanzee (Šimpanz)
A chimp lies on **his** back in a hammock of branches. One arm is behind **his** head, the other is crooked at the elbow.
National Geographic April 2008, p. 144

Šimpanz leží na zádech ve visutém lůžku z větví a kotníkem jedné nohy se opírá o ohnuté koleno druhé nohy.
National Geographic Duben 2008, str. 144

The following English version contains only English contexts without gender markers.

4.6.1 The use of gender markers by Czech students

Czech university students of English were asked to fill in personal and possessive pronouns into the following sentences:

Acoel
An acoel, a tiny flatworm without a gut, inhabits stony shallows. _____ noxious skin puts off predators, and _____ head can regenerate in days if severed.

Caterpillar
In one-twelfth of a second the caterpillar whips back and ensnares the termite, then gobbles _____ meal. Belly full, _____ hangs languidly and finishes off the last bits.

Worm
This reef dweller protects _____ six-inch body in a sand-and-mucus tube while _____ appendages drift.

Jellyfish
This jellyfish was dubbed „big ugly“ when _____ was discovered. _____ can grow to more than three feet.

Mantid
_____ seems almost human, this mantid I found in West Africa. _____ has such alert eyes, and _____ head tilts to follow me.

Salmon
After defending _____ nest for as long as _____ can swim upright, a salmon drifts back into an eddy to die.

Scorpion
In some, like this Arizona species , the scorpion stands guard over _____ eggs and later _____ newborns.

Albatross
When the parent arrives, _____ and the chick excitedly position their bills cross-wise. An adult may spend 15 minutes ashore feeding _____ youngster.

Crane
A red-crowned dancer bows _____ head and flaps _____ wings.

Wolf of Ethiopia
Remember that young wolf who's babysitting the pups at the den. _____ 's at the age where it's time for _____ to move out and find _____ own pack.

Kinkajou
Kinks usually bear one pup every year or two, _____ for as long as four months.

Honey badger
One night we saw a young badger collapse. _____ 'd been struck in the face by a puff adder just before _____ bit its head off.

Lynx
A lynch in Canada, _____ right forefoot caught in a snare, awaits _____ fate.

Hyena
This young hyena has no intention of sharing. _____ could lose _____ lunch entirely.

Leopard
When the cub was five months old, _____ mother brought _____ a live baby impala.

Moose
_____ no longer shook _____ antlers at his assailants.

Snow Fox
Heavy rain may have prompted _____ decision, or _____ may have soust a cleaner site.

Bear
A week or two out of hibernation, _____ s spending the day filling _____ belly on the first lush green of spring.

Panda
Tai-Shan will stay at the National Zoo with _____ mother until _____ is two years old.

Orca
Rather, _____ is herding the fish until _____ calf joins the hunt.

Chimpanzee
A chimp lies on _____ back in a hammock of branches. One arm is behind _____ head, the other is crooked at the elbow.

The total number of respondents for this analysis was 127 students, 106 females and 21 males, all of them studying at the University of South Bohemia, Pedagogical Faculty and Faculty of Arts.

The students' choices are recorded in the following table, which has been divided into five main columns. The first column contains the list of animals. The second shows the total number of students and is divided according to their sex. The next three columns contain the number of the pronouns which students used to refer to the animals.

Table 4.38: Survey results

Animal	Number of respondents			Masculine			Feminine			Neuter		
	F	M	T	F	M	T	F	M	T	F	M	T
Acoel	106	21	127	17	3	20	11	2	13	78	16	94
Caterpillar	106	21	127	11	2	13	35	7	42	60	12	72
Worm	106	21	127	49	10	59	1	0	1	56	11	67
Jellyfish	106	21	127	6	1	7	44	9	53	56	11	67
Mantid	106	21	127	6	1	7	38	8	46	62	12	74
Salmon	106	21	127	66	13	79	1	0	1	39	8	47
Scorpion	106	21	127	55	11	66	2	0	2	49	10	59
Albatross	106	21	127	71	14	85	1	0	1	34	7	41
Crane	106	21	127	23	5	28	6	1	7	77	15	92
Wolf of Ethiopia	106	21	127	82	16	98	5	1	6	19	4	23
Kinkajou	106	21	127	26	4	30	2	1	3	78	16	94
Honey badger	106	21	127	22	4	26	0	0	0	84	17	101
Lynx	106	21	127	72	14	86	11	2	13	23	5	28

Animal	Number of respondents			Masculine			Feminine			Neuter		
	F	M	T	F	M	T	F	M	T	F	M	T
Hyena	106	21	127	6	1	7	25	5	30	75	15	90
Leopard	106	21	127	13	3	16	4	1	5	89	17	106
Moose	106	21	127	72	15	87	2	0	2	32	6	38
Snow fox	106	21	127	27	5	32	63	13	76	16	3	19
Bear	106	21	127	90	18	108	1	0	1	15	3	18
Panda	106	21	127	22	4	26	33	7	40	51	10	61
Orca	106	21	127	6	1	7	60	12	72	40	8	48
Chimpanzee	106	21	127	84	16	100	3	1	4	19	4	23
Total						**987**			**418**			**1262**

Data analysis:

For the purpose of this analysis, both genders, the gender of the animal in English as well as the gender assignment in Czech, are important.

The Czech translations include 21 animal names:

In Czech

thirteen animals are **masculine**

- rournatec (worm), losos (salmon), škorpion (scorpion), albatros (albatross), jeřáb (crane), vlk (wolf), medojed (honeybadger), kinkadžů (kinkajou), rys (lynx), leopard (leopard), los (moose), medvěd (bear), šimpanz (chimpanzee)

eight animals are **feminine**

- ploštěnka (acoel), housenka (caterpillar), medůza (jellyfish), kudlanka (mantid), hyena (hyena), liška (fox), panda (panda), kosatka (orca)

There is no adult animal with neuter gender reference in Czech, but according to the contexts above we may consider the crane, honey badger, kinkajou, hyena, leopard and panda as youngsters.

All listed animals required an explicit gender reference within the excerpted contexts. I did not give the students any instructions for filling in the appropriate pronouns. I have only made sure they understood all the animal names.

Surprisingly, I have found no discrepancies in the students' answers. They used only one type of pronoun- either personal or possessive- in their short contexts.

The following charts show the correlation between the students' responses and the gender of the animal in Czech:

Acoel – **feminine** in Czech

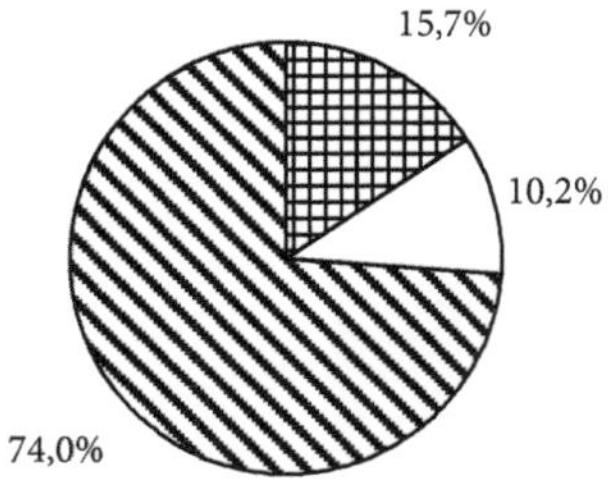

Caterpillar – **feminine** in Czech

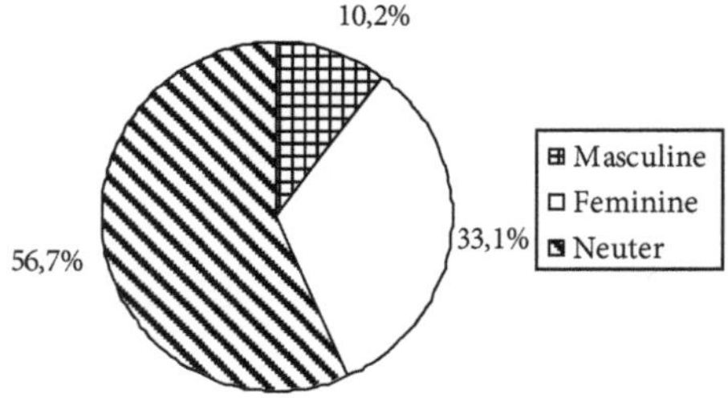

Worm – **masculine** in Czech

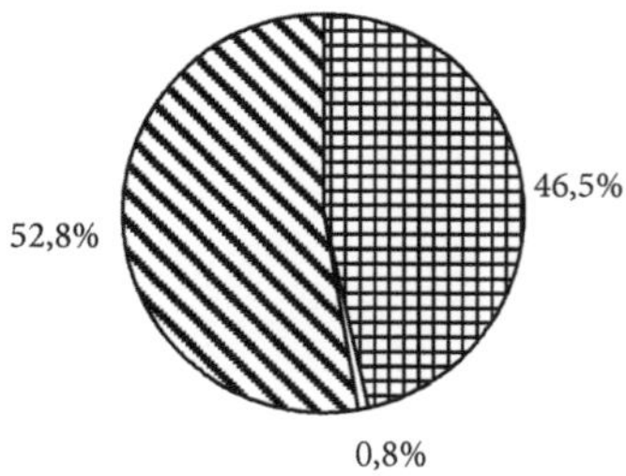

Jellyfish – **feminine** in Czech

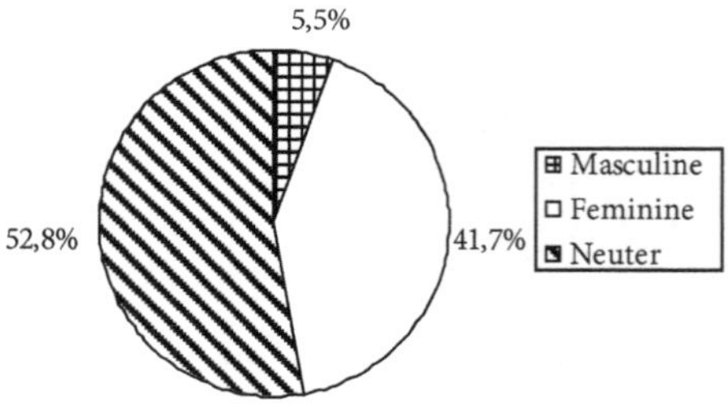

Mantid – **feminine** in Czech

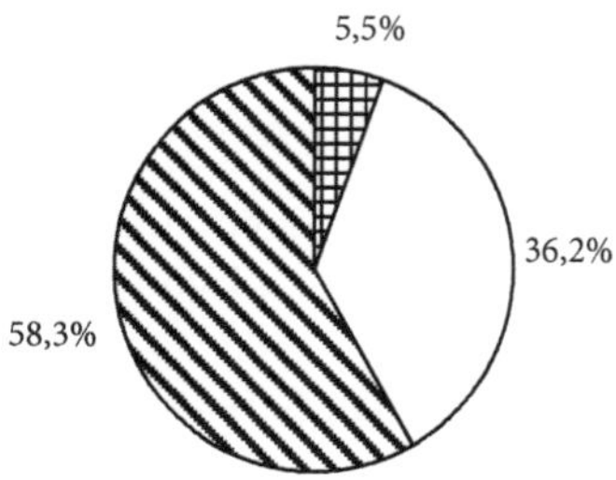

Salmon – **masculine** in Czech

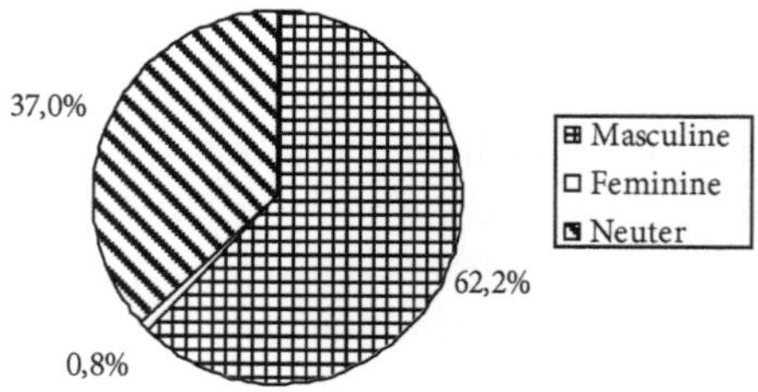

Scorpion - **masculine** in Czech

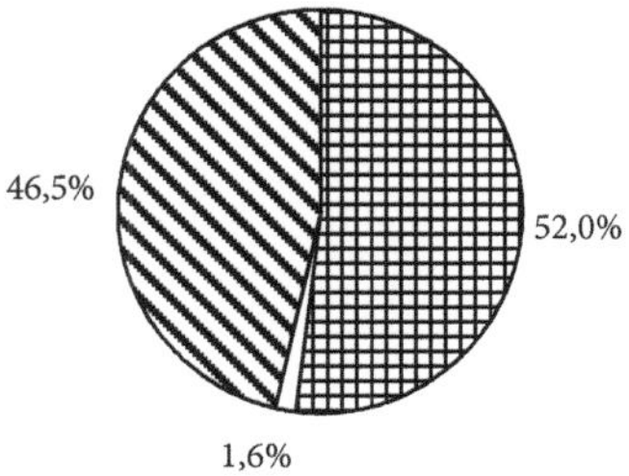

Albatross - **masculine** in Czech

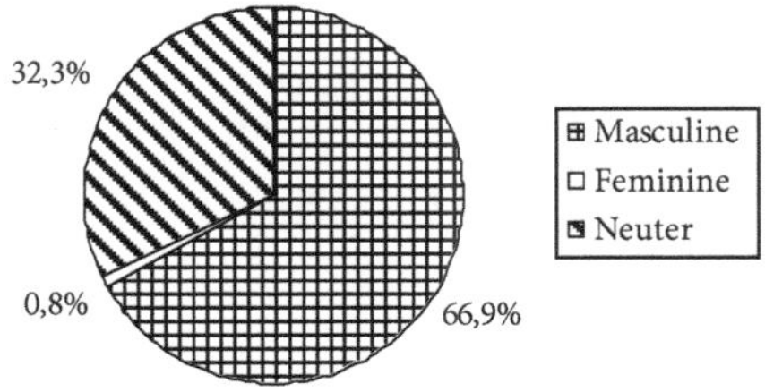

Crane - **masculine** in Czech

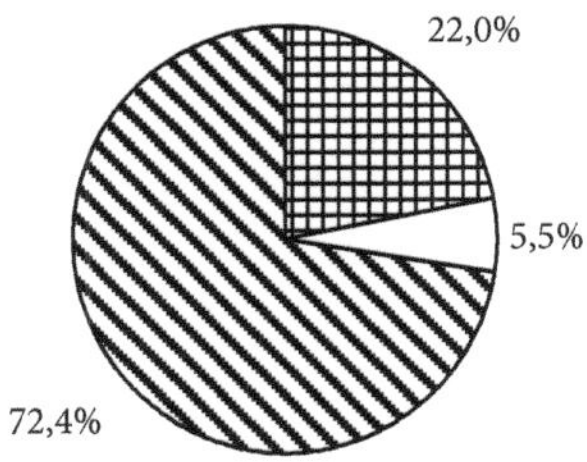

Wolf - **masculine** in Czech

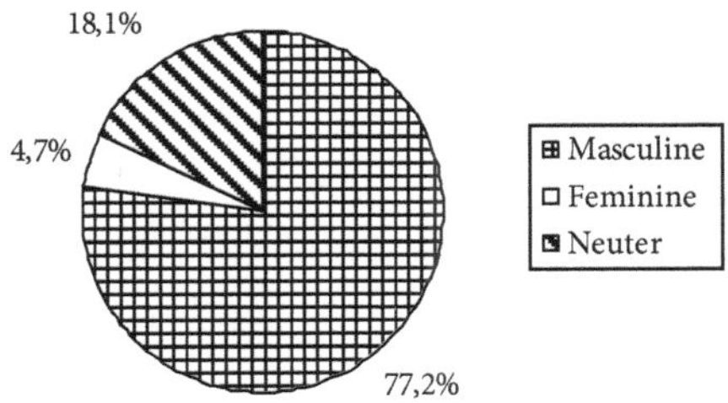

Kinkajou - **masculine** in Czech

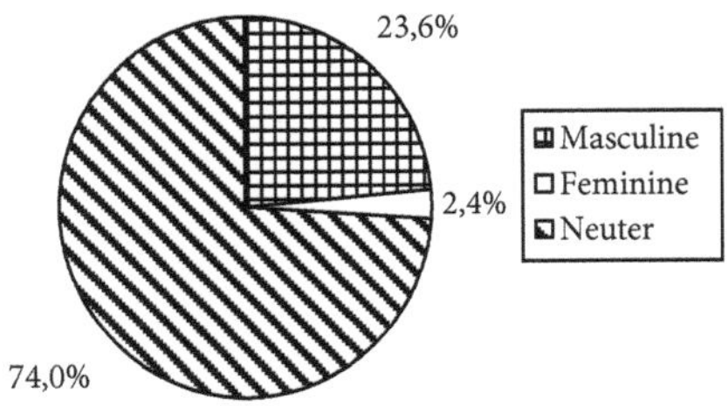

Honey badger – **masculine** in Czech

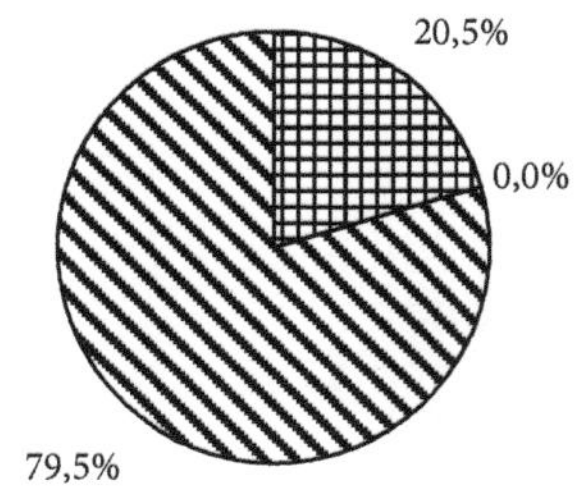

Lynx – **masculine** in Czech

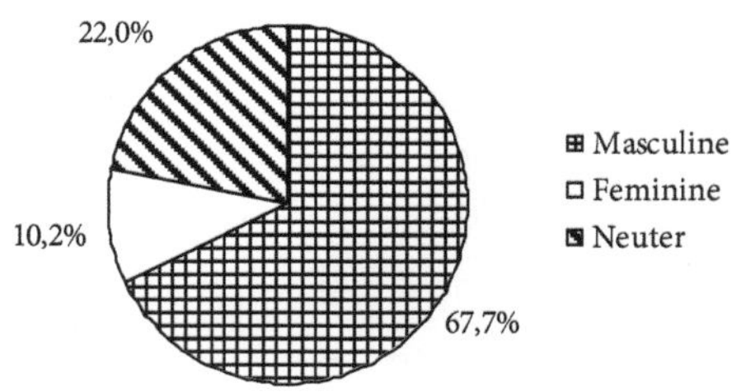

Hyena – **feminine** in Czech

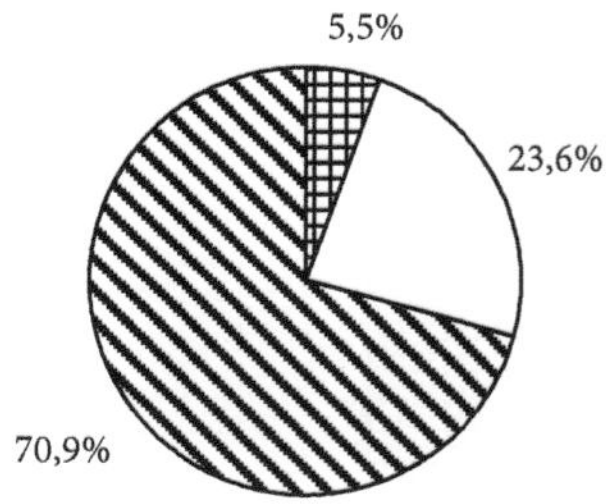

Leopard – **masculine** in Czech

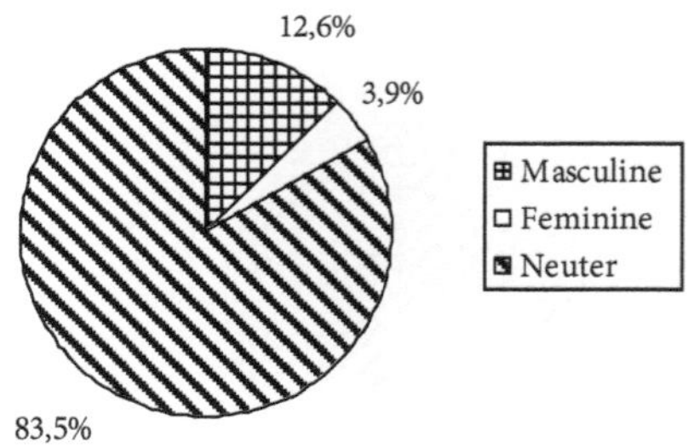

Moose – **masculine** in Czech

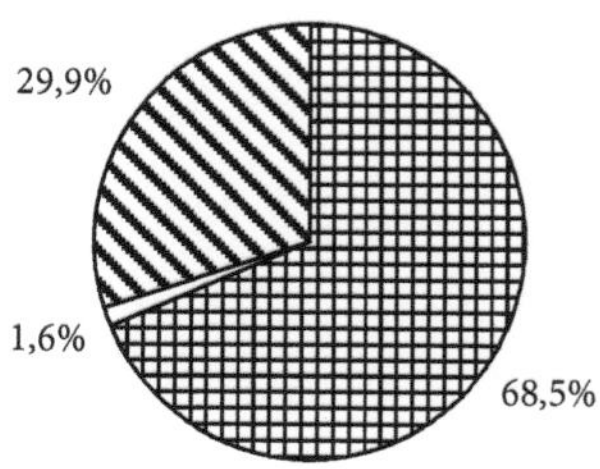

Fox – **feminine** in Czech

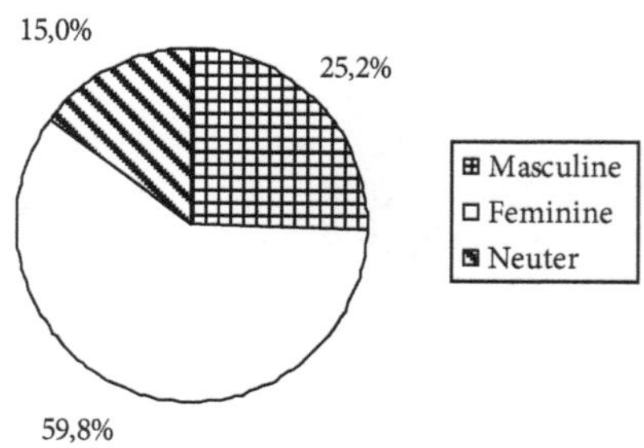

Bear – **masculine** in Czech

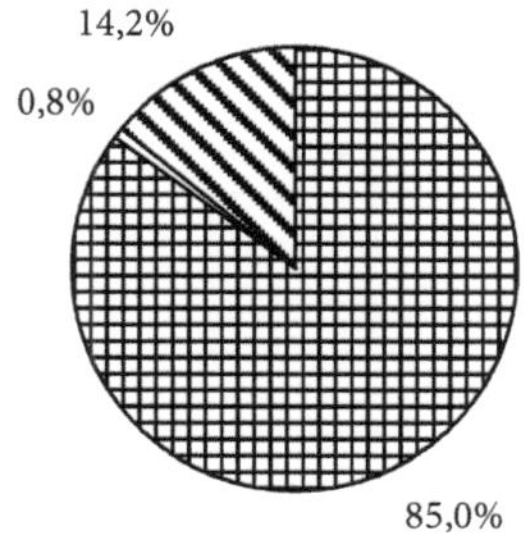

Panda – **feminine** in Czech

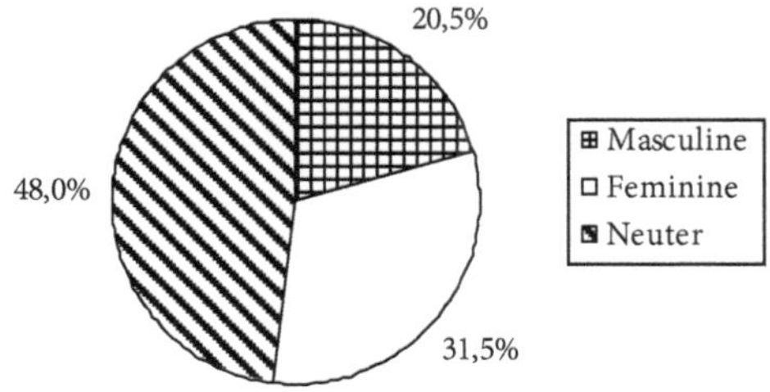

Orca – **feminine** in Czech

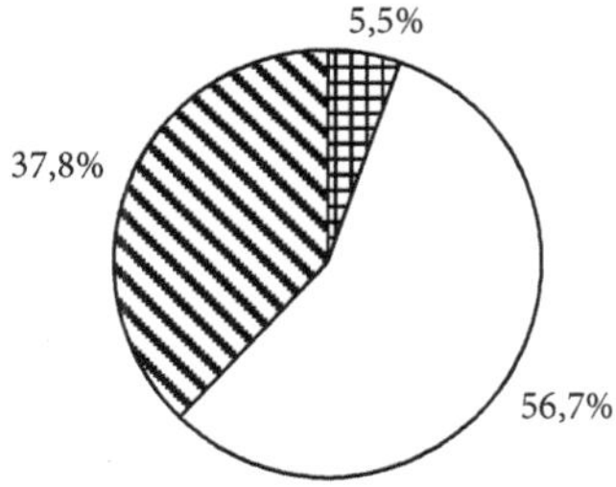

Chimpanzee – **masculine** in Czech

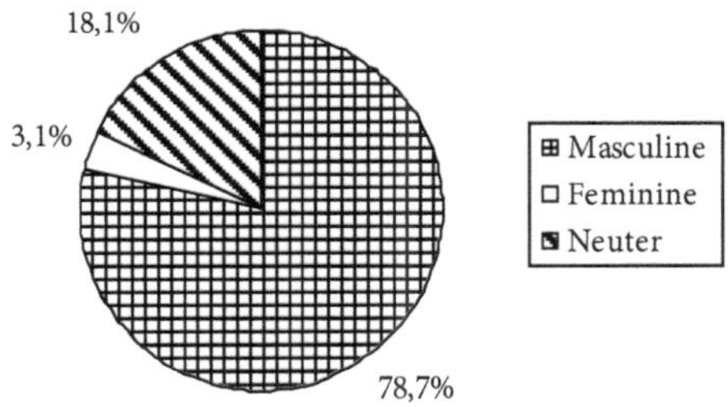

In the following tables the animal nouns have been divided according to their gender assignment in Czech. The aim is to compare the gender in Czech and the gender markers the students used in the sentences excerpted above.

Nouns that are **masculine in Czech** were referred to in English as follows:

Table 4.39: Animals – masculine in Czech

	M	F	N
Worm	**46.5**	0.8	52.8
Salmon	**62.2**	0.8	37.0
Scorpion	**52.0**	1.6	46.5
Albatross	**66.9**	0.8	32.3
Crane	**22.0**	5.5	72.4

	M	F	N
Wolf	**77.2**	4.7	18.1
Kinkajou	**23.6**	2.4	74.0
Honey badger	**20.5**	0.0	79.5
Lynx	**67.7**	10.2	22.0
Leopard	**12.6**	3.9	83.5
Moose	**68.5**	1.6	29.9
Bear	**85.0**	0.8	14.2
Chimpanzee	**78.7**	3.1	18.1

Nouns that are **feminine in Czech** were referred to English as follows:

Table 4.40: Animals – feminine in Czech

	F	M	N
Acoel	**10.2**	15.7	74.0
Caterpillar	**33.1**	10.2	56.7
Jellyfish	**41.7**	5.5	52.8
Mantid	**36.2**	5.5	58.3
Fox	**59.8**	25.2	15.0
Panda	**31.5**	20.5	48.0
Orca	**56.7**	5.5	37.8

The following table lists the youngsters and their gender assignments in English. The appropriate adult animals are added in brackets:

Table 4.41: Youngsters in English

	N	M	F
Young (crane)	**72.4**	22.0	5.5
Young (honey badger)	**79.5**	20.5	0.0
Pup (kinkajou)	**74.0**	23.6	2.4
Young (hyena)	**70.9**	5.5	23.6
Cub (leopard)	**83.5**	12.6	3.9
Cub named Tai-Shan (panda)	**48.0**	20.5	31.5

The three tables show a strong influence of Czech grammatical gender on the gender assignment in English. As can be seen from the following three charts, the Czech students have assigned the gender to the animals according to their Czech equivalents in the majority of sentences. Surprisingly, the least influence of Czech gender assignment in English texts can be seen in the youngsters, grouped in the last above table. The collected data have been recorded in the relevant tables according to the sex of the respondents. In my analysis I have not found any differences between male and female users of English.

4.6.2 The use of gender markers by native speakers

To compare gender assignment in animals by Czechs and by native speakers of English, respectively, 12 native speakers -lecturers of English at South Bohemian University- were asked to complete the same sentences. The instructions were the same. The below table shows their choices:

Table 4.42: Native speakers' responses

Animal	Number of respondents	Masculine	Feminine	Neuter
Acoel	12			12
Caterpillar	12			12
Worm	12	1		11
Jellyfish	12	1		11
Mantid	12			12
Salmon	12	2	3	7
Scorpion	12		9	3
Albatross	12	4	4	4
Crane	12	3		9
Wolf of Ethiopia	12		11	1
Kinkajou	12			12
Honey badger	12	3		9
Lynx	12	5		7
Hyena	12	1		11
Leopard	12	5		7
Moose	12	9		3
Snow fox	12		3	9

Animal	Number of respondents	Masculine	Feminine	Neuter
Bear	12	2		10
Panda	12	1	1	10
Orca	12		12	
Chimpanzee	12	3		9
Total	12	**40**	**43**	**169**

The above data show that native speakers of English tend to assign the majority of animals to by neuter.

Final charts and analysis:

Masculine animal nouns in Czech and their gender assignment in English (pronouns referring to thirteen animals):

Figure 4.40: Gender assignment – Masculine nouns in Czech.

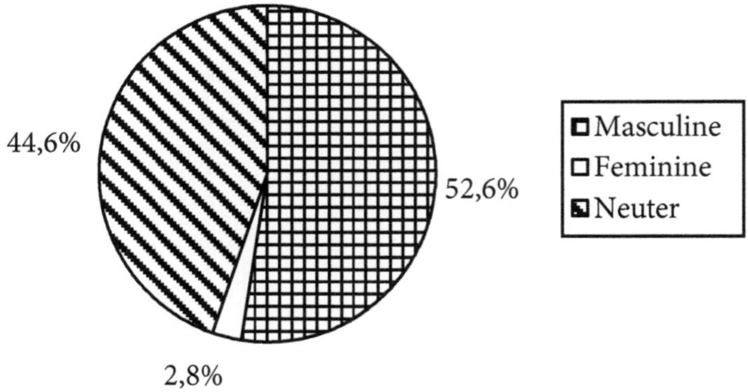

Feminine animal nouns in Czech and their gender assignment in English (pronouns referring to seven animals):

Figure 4.41: Gender assignment – Feminine nouns in Czech.

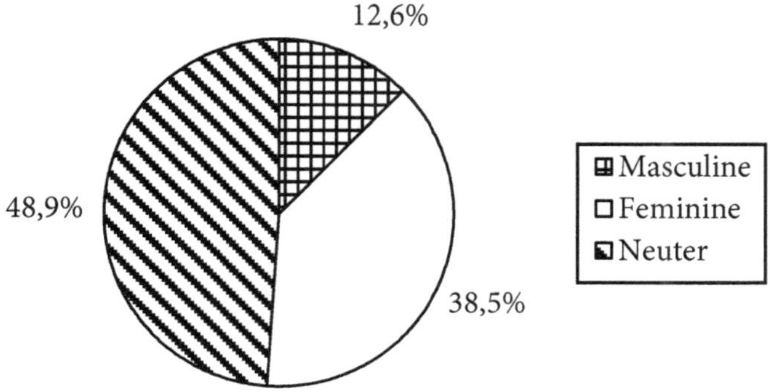

Six youngsters in English contexts and their gender assignment:

Figure 4.42: Gender assignment – Youngsters in English.

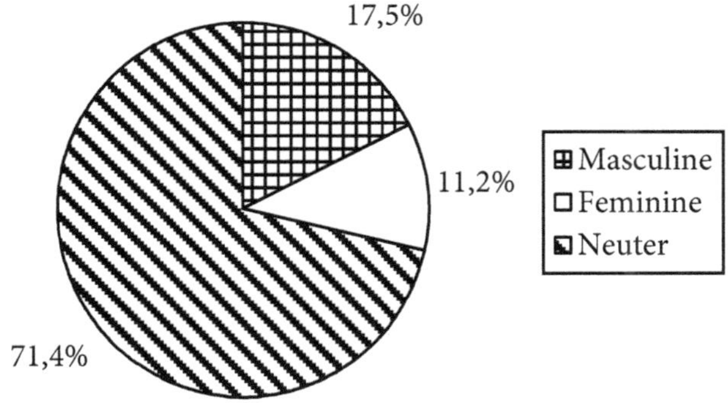

As can be seen from the above final charts, the influence of Czech with the formal assignment system is very strong in assigning gender in English.

Masculine nouns in Czech were referred to as masculine in English in 52.6%
Feminine nouns in Czech were referred to as feminine in English in 38.5%
Youngsters were referred to as neuter in 71.4%.

In the case of low animals such as acoel, caterpillar, worm, jellyfish and mantid, students used mainly neuter pronouns. This fact can be based on the students' theoretical knowledge of classification of nouns referring to animals. According to most grammars, these species belong to "low" and "less familiar" animals and thus are referred to as ***it***. The same situation is manifested in the group of youngsters. The majority of them were referred to by ***it***. The explanation might be based on the theoretical knowledge of assigning gender markers in youngsters in English.

On the other hand, the group of higher animals shows a very strong influence of Czech. The scorpion, albatross, wolf, lynx, moose, bear and chimpanzee, which are masculine in Czech, have been mostly referred to by masculine pronouns in English sentences.

4.6.3 The use of gender markers by Czech students and native speakers

Czech students' perceptions Native speakers' perceptions

Figure 4.43: Comparative charts – speakers' perceptions.

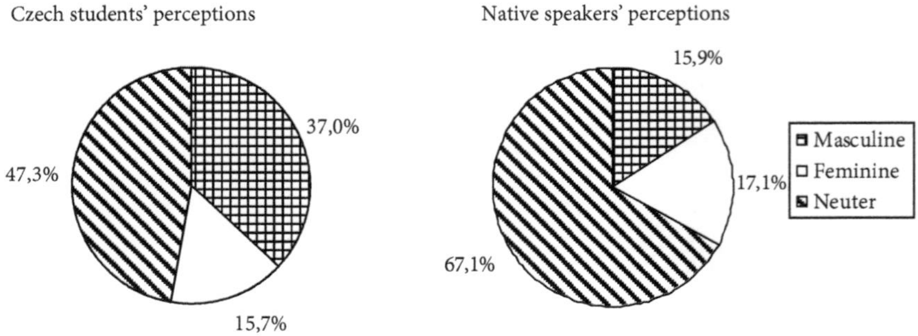

The comparison of gender marking by Czech students and native speakers has revealed substantial differences. As can be seen from the above comparative pair of charts, Czech speakers of English prefer masculine when assigning gender to animals. The results of this analysis show a very strong influence of gender assignment system in Czech on Czech users of English when assigning gender in English. The above charts show the tendency to apply the Czech system of gender assignment when assigning gender in English. The majority of Czech students assigned the same gender to animal nouns as is used in Czech.

Native speakers of English used predominantly neuter when referring to 21 animal species. This fact could be explained by general usage of neuter when referring to animals in English.

Above all, lecturers and students included in this analysis are not zoologists. For lay users it might be difficult to distinguish between familiar and less familiar animals. They could not have any knowledge about the sex of such unknown animals as acoel, honey badger or kinkajou. Consequently, all speakers of English were uncertain and hesitant about assigning gender to these less familiar animals and thus have used predominantly the neuter.

Another possible explanation might be contained in Huddleston's definition (2002: 489): *"**it** does not mean "neither male nor female". In many cases the neuter pronoun is used because the speaker doesn't know what the sex is."*

The next factor that might have influenced all respondents of this analysis was that of emotion and personal involvement, especially in animals which rank high in the kingdom of wild animals. The use of feminine and masculine especially in high animals could be related to the greater degree of interest in or empathy with the animal. One of the examples that can support this assumption is gender assignment in orca. The majority of respondents referred orca as ***she*** (56.7% of Czech students and 100% English lecturers). Their choice could be supported by the context in the sentence, referring to maternity care.

Even in my research based on *National Geographic* articles orca and whale species are viewed predominantly as ***she*** (in 45.6% of all examples retrieved).

5. Summary

This thesis aims to be a contribution to gender studies by providing the answer to the core question: **What are the criteria of assigning gender to animal nouns?**

This question is topical in view of the fact that in present-day English, gender is a marginal grammatical category and therefore recedes into the background in descriptions of today's English. Grammatical gender is a rare topic of research and scientific discussions or papers.

To establish the theoretical background for my research, it was first necessary to define the term "gender" because it causes a lot of misunderstanding, often being used to define different concepts. When the research focus is on gender as manifested in language, the usual objectives are to discuss the differences between male and female use of language, or the differences in the representation of women and men in a language.

Chapter 2 is thus concerned with a brief description of the history of gender exploration and an overview of gender in English. As my priority was not to study the history of gender, I have only discussed these topics briefly in order to set the scene for the chapters dealing with recent descriptions of gender assignment in animal nouns in various reference books. Although the respective authors base their studies on a variety of different corpora, such as children's literature, web pages devoted to animal keeping and breeding and students' use of personal pronouns in tag questions, the major results of my research are rather unexpected. As all of the authors found a surprisingly high number of "human" pronouns used to refer to animals, they claim that masculine is the prevailing gender. The findings offered by valuable authors can be summarized as follows:

- The choice of pronouns referring to animals is affected by a number of emotive factors and by the degree of personal involvement.
- The cut-off point within the class of animals differs from speaker (or author) to speaker depending on their professions, environment, etc.
- When referring to animal nouns, neuter pronouns are the least frequently occurring forms, while the masculine dominates.

These results were crucial for my further research, but highly unexpected in light of traditional grammar descriptions. In subchapter 3.2. I summarized the grammarians' views on gender in present-day English. I have compared various modern reference grammars and grammar practice books. To my knowledge, the most comprehensive explanations concerning the gender of animals are found in

Huddleston and Pullum's *Cambridge Grammar of the English Language* (2002). In contrast, most other grammars do not pay appropriate attention to gender assignment in animals. The authors focus mainly on terminological differentiation and professional language use and do not mention spoken language referring to animals. In discussing the gender of animals, more attention in grammar books is given to the word-formation processes or lexical items distinguishing heteronymous animal pairs than to the ways of assigning gender to unmarked nouns. Most grammars, not very clearly, divide animals into familiar and less familiar, or higher and lower animals. Only more comprehensive grammars account for the role of other factors in assigning gender, such as the degree of interest in and empathy with the animal, its position in the animal kingdom, as well as emotional involvement, tradition and convention.

Subchapter 3.3 contains an overview of gender in animals in selected world's languages, the basic terminology and classifications according to two major systems for assigning gender: semantic and formal. The focus was on assigning gender to animal nouns, both in languages with strict semantic or predominantly semantic systems, and those with formal systems. The formal systems have been illustrated on the example of German, French, Russian and Bantu.

Subchapter 3.3. also deals with assigning gender to animals in Czech. I have included examples in both Czech and English to demonstrate the fact that the explanation of gender in animal nouns is not straightforward. The above observations suggest that gender assignment in English is based on a variety of factors that go beyond linguistics. The word "gender" cannot be replaced by the Czech term "rod". Since this Czech expression does not include the social and cultural aspects. We have to distinguish gender (rod) as a grammatical and biological category and gender as a sociological category.

A detailed analysis of gender assignment to animal names based on the main corpus and two subcorpora of this study is presented in Chapter 4.

To be able to offer a more complex view of the above aspects of gender and language, the following set of research questions were defined at the very beginning of my analysis:

- ***How is gender assigned to animal nouns in English?***
- ***Are there any differences in assigning gender to animal nouns in the corpora explored?***
- ***How do Czech speakers assign gender to animal nouns in English?***
- ***To what extent do the differences between gender assignment systems in English and in Czech influence Czech speakers of English in their choice?***

In the introductory part of Chapter 4 I justify the compilation of my corpus and data, which can be found in the appendices. The objectives of this analysis were to identify the rules and principles of gender assignment to animal nouns retrieved from the articles. I divided all the examples retrieved into classes and species according to their biological classification. The results for each animal group and its referring pronouns are presented graphically in tables and charts, and verbally, in the form of comments and quotations.

The main corpus for my analysis was retrieved from the English edition of *National Geographic*. As a life-member of *National Geographic Society* since 1979 I own the 36 years' collection of this magazine. Taking into account my degree in biology, the topic of this thesis seemed to provide an interesting field for exploration in gender studies.

The analysis of all zoological articles revealed the prevalence of feminine gender markers in all animal species in question. The "lowest" animals with gender distinctions found in the corpus were insect species. On the other hand the "highest" animals from the point of view of zoological hierarchy were apes. Out of 2,753 pronominal pronouns 46.6% of all pronouns were feminine, 32.9% masculine and 20.5% neuter. Surprisingly, the highest occurrence of feminine pronouns was recorded in ant species. This fact could be explained by the specific social life cycles of these insects. The other species with the dominance of feminine in their gender assignment were bears (65.8%), lions (58.9%), tigers (53.2%), reptiles and amphibians (48.7%) and whales (45.6%). Feminine was predominantly used in situations describing maternal care and reproduction. Masculine prevalence occurred in seahorse species (74.4%), exotic birds (56.7%), birds of prey (53.1%), leopards (52.9%) and apes (49.5%). The predominance of masculine gender is probably the result of extensive descriptions of behaviours, life cycles and reproduction processes, in which the biological gender plays a significant role.

The neuter proved to be a surprisingly marginal category. It is the most frequent gender when referring to marine microfauna (75%), tropical fishes (73.9%) and shark species (53.7%). The possible explanation of the neuter gender prevalence might be the wide variety of microfauna and fish species. The authors had no need to indicate the sex of the animals. General descriptions of life cycle prevailed. On the other hand, neuter was totally absent from descriptions of tiger, elephant and seahorse species. The explanation of this fact is simple; strong personal involvement in the case of the endangered tiger and elephant and the unique reproductive process of the seahorse.

Concluding from the authentic data, the assignment of gender to animal nouns in *National Geographic* articles does not fully follow the rules of semantic assignment system. Even though the majority of animal species are assigned to genders according to their biological sex, there are other important factors that influence the gender marking. The choice of the pronoun also depends on the social roles of the animals and their status in "the animal kingdom". Many of the animals retrieved from *National Geographic* represent threatened species and are among the most recognizable and popular of the world's megafauna. A good example could be "the big cat" species, such as jaguar, leopard and tiger, "the most charismatic animal" (cited from *National Geographic* 1984: 763) in my corpus. Tigers are personalised, have names and are referred to entirely by *he* or *she*. Huddleston and Pullum's statement that *he* and *she* are "more likely with the creatures ranked high in the kingdom of wild animals (e.g. *lions, tigers, elephants*)" (Huddleston et al. 2000: 489) is very much in agreement with my own observations. The distinction between higher and lower animals as it occurs in *National Geographic* does not necessarily correspond to the biological classification, so lower animals are very often viewed as higher animals.

Apart from their grammatical functions, the pronouns imply a sociological function. Although the corpus was based on zoological articles and the biological sex of the individuals concerned played an important role, the findings of my research confirm Vachek's observation that "feminine or masculine reference signals some interest of the speaker in the individual referred to. There may be some interference of emotional factors" (Vachek 1964: 191).

Comparing my conclusions with the recent works surveyed in the introductory part, feminine prevails in my corpus while all the other authors contend that the masculine gender is dominant. Based both on my own study and on other recent studies in the area of assigning gender to animals, the following conclusion can be drawn, providing an answer to the original question: ***"How is gender assigned to animal nouns in English?"*** When referring to animal nouns, neuter pronouns are the least frequently occurring forms. Surprisingly, a significant role in the choice of pronouns referring to animals is played by emotional factors, personal involvement, biological classification, tradition and convention.

To sum up, the common claim that "the pronoun ***it*** is used to refer to animals predominantly in scientific or non-literary contexts where there is no need to express personal or social attitudes" is not universally valid.

Chapter 4.3. is concerned with the study of a subcorpus based on 276 original research papers taken from recent three years of Journal of Zoology. The aim of this partial research was the identification and analysis of gender distinction – if any – in animals retrieved from the research papers. The focus was primarily on the occurrence of masculine and feminine pronouns. I listed 217 animal species with 844 gender references. The final counts revealed a strong preference for the neuter, with 73.9 % of animals assigned exclusively to neuter. On the other hand, the feminine gender assignment was surprisingly high. A possible explanation of this fact could be that the animals analysed are mostly females with their young, a category most interesting for researchers. To sum up, in all the selected examples the assignment of gender to animal nouns was primarily based on their natural biological sex. A very important factor that influenced gender assignment within this corpus was that of emotion and personal involvement of the researchers.

Detailed comparative analyses based on the subcorpora of *National Geographic* and *Journal of Zoology* were described in Chapter 4.4.

My aim was to ascertain the differences of gender assignment in identical animal species taken from the two different sources. I analysed and compared a set of 16 species of animals, starting with the lowest insect species and finishing with big mammals. Although the number of the pronouns retrieved is higher in *National Geographic* articles (752 pronouns retrieved from *National Geographic* and 413 pronouns taken from *Journal of Zoology*), the distribution of all masculine, feminine and neuter gender markers in both sources was surprisingly balanced. In both corpora feminine and neuter gender markers dominated. The most illustrative example supporting this finding could be the gender assignment in the "lowest" animal of this comparative analysis – the praying mantis. As shown by two comparative charts, the proportions of masculine pronouns in this animal are the same and the occurrence of feminine and neuter markers are very similar. My explanation of this fact is straightforward. The reproductive process and courtship strategies of this animal are very specific, thus all the articles taken from the different corpora had nearly the same structure and gender references.

To conclude, all the species yielded a high number of masculine and feminine gender markers in both subcorpora, which is very surprising, especially in research papers. I would not expect the presence of masculine and feminine gender and personal involvement in animals in such scientific research material as represented by *Journal of Zoology*. Consequently, I could answer the second research question defined in the introduction ***"Are there any differences in assigning gender to animal nouns in the corpora explored?"***

I did not find any substantial differences within the two subcorpora. The main criteria for gender distinctions in both studied materials are very similar:

- The description of mating and courtship processes
- The description of motherhood and parental care
- The close relationship between the researchers and the animal in long-term studies

Chapter 4.5. was devoted to animal youngsters. My aim was to answer the question: *How is the gender of juvenile animals distributed?* The initial hypothesis was that gender assignment to juvenile animals would manifest differences from that in adults. I collected two sets of data, retrieved predominantly from *National Geographic*. The second subcorpus represents *Journal of Zoology*. The results showed the tendency to use predominantly masculine reference in *National Geographic* (47.3%). The distribution of gender markers in the corpus containing research papers was different, with 83.8% of the pronouns referring to youngsters neuter, and 12.5% feminine.

To sum up, even if there was a disproportion between the number of excerpted gender markers and youngsters taken from the corpora explored, the general rules of gender assignment in juveniles seemed to be very similar. The gender distinctions in youngsters were used predominantly in the descriptions of

- parental care and family relationships
- research and observation related to a particular youngster.

Although most grammars of both modern and earlier stages of English either ignore the topic or maintain that the appropriate pronoun to use when referring to youngsters is ***it***, the actual language use identified by my research is quite different.

Although grammatical gender and natural sex play a very important role in assigning gender to youngsters, the emotional involvement of the authors or researchers is apparent in the majority of the examples retrieved.

Chapter 4.6. focused on the differences in gender assignment by English and Czech speakers. I compared Czech students' and native speakers' use of gender markers based on their perceptions of the respective animals. The speakers were asked to assign gender to 21 species of animals in short contexts taken from *National Geographic*. This comparison revealed very different gender markings. Czech speakers of English preferred masculine, whereas native speakers used predominantly neuter. This fact can be explained by appealing to Vachek's observations, namely, that "neuter reference is commonly resorted to for the simple reason that the sex of the concerned individual is either unknown or irrelevant" (Vachek 1964:191).

The summary of the data provides an answer to the last research question "***To what extent do the differences between gender assignment systems in English and Czech influence Czech speakers of English?***" as follows: The differences between the gender assignment systems in English and Czech have a very strong influence on Czech users of English. My analysis showed the tendency to apply the Czech system of gender assignment when assigning gender in English. Both groups of respondents might be influenced by general usage of neuter when referring to animals in English. **All in all, gender assignment in animals is the result of an interplay of different factors, not all of them purely linguistic.** For someone who is not a zoologist it might be difficult to distinguish the sex of such less well-known animals as acoel or kinkajou. On the other hand, lay users will probably restrict gender distinction only to pets, whereas a strange creature such as honey badger, which the speaker may never have seen, will be ***it***. Another influential aspect could be that of **emotion** and **personal involvement**, especially in referring to "higher" animals.

List of Figures

List of Tables

Bibliography

Alexander, L.G. (1991) *Longman English Grammar Practice.* London: Longman.

Alexander, L.G. (1998) *Longman English Grammar.* London: Longman.

Anderson, J. (1792) "Grammatical Disquisitions." *The Bee* 11: 120–30; 193–204; 240–50.

Asher, R.A. (ed) (1994) *The Encyclopedia of Language and Linguistics.* 10 volumes. Oxford: Pergamon Press.

Beaumont, D. and C. Granger (1989) *The Heinemann English Grammar.* Oxford: Heinemann.

Biber, D., Johansson, S., Leech, G., Conrad, S. and E. Finegan (1999) *Longman Grammar of Spoken and Written English.* London: Longman.

Biber, D., Johansson, S., Leech, G., Conrad, S. and E. Finegan (1999) *Longman Grammar of Spoken and Written English.* London: Longman.

Biber, D., Conrad, S. and R. Reppen (1998) *Corpus Linguistics. Investigating Language Structure and Use.* Cambridge: Cambridge University Press.

Blair, H. (1783) *Lectures on Rhetoric and Belles Lettres.* London: W. Strahan.

Bodine, A. (1975) "Androcentrism in Prescriptive Grammar: Singular "They", Sex- Indefinite "He", and "He or She"." *Language in Society* 4: 129–46.

Bodine, A. (1998) "Androcentrism in Prescriptive Grammar: Singular "They", Sex- Indefinite "He", and "He or She"." In *The Feminist Critique of Language,* ed. D. Cameron, 124–138. London: Routledge.

Brinton, L.J. (2000) *The Structure of Modern English.* Amsterdam/Philadelphia: Benjamins.

Brown, G. [1851] (1880) *Grammar of English Grammars.* 10th ed. New York.

Cameron, D. (1994) "Verbal Hygiene for Women: Linguistics Misapplied." *Applied Linguistics* 15: 382–398.

Cameron, D. (1995) *Verbal Hygiene.* London: Routledge.

Cameron, D. Frazer, E., Harvey, P., Rampton, M.B.H. and K. Richardson (1992) *Researching Language: Issues of Power and Methods.* London: Routledge.

Cameron, D. (1994) *Feminism and Linguistic Theory.* London: Macmillan.

Coates, J. and D. Cameron (1988) *Women in Their Speech Communities.* London: Longman.

Coates, J. (1993) *Women, Men and Language. A Sociolinguistic Account of Gender Differences in Language.* London: Longman.

Corbett, G. and N. M. Fraser (2000) "Default genders." In *Gender in Grammar and Cognition*, eds. B. Unterbeck et al., 55–97. Berlin: de Gruyter.

Corbett, G. (1991) *Gender.* Cambridge: Cambridge University Press.

Crawford, M. (1995) *Talking Difference. On Gender and Language.* London: Sage Publications.

Croft, W. (1990) *Typology and Universals.* Cambridge: Cambridge University Press.

Crombie, A. [1802] (1830) *The Etymology and Syntax of the English Language Explained and Illustrated.* 3d ed. London: J. Taylor.

Crystal, D. (1995) *The Cambridge Encyclopedia of the English Language.* Cambridge: Cambridge University Press.

Curme, G. O. (1962) *English Grammar.* New York: Bames & Noble.

Curzan, A. (2000) "Gender Categories in Early English Grammars: Their Message to the Modem Grammarian." In *Gender in Grammar and Cognition*, eds. B. Unterbeck et al., 561–576. Berlin: de Gruyter.

Dekeyser, X. (1980) "The Diachrony of Gender Systems in English and Dutch." In *Historical Morphology*, ed. J. Fisiak, 97–111. The Hague: Mouton.

Dickins, T. (2001) "Gender Differentiation and the Asymmetrical Use of Animate Nouns in Contemporary Czech." *The Slavonic and East European Review* 79: 212–247.

Dušková, L. (1988) *Mluvnice současné angličtiny na pozadí češtiny.* Praha: Academia.

Dušková, L. (1965) "On Some Disputed Points in the Use of Pronouns in Present Day English." *Philologia Pragensia* 213: 163–170.

Encyclopedia Britannica (1999) "Multimedia Edition." On CD-ROM.

Fennell, B. (2001) *A History of English. A Sociolinguislic Approach.* Oxford: Blackwell.

Greenbaum, S. and R. Quirk (1993) *A Student' s Grammar of the English Language.* London: Longman.

Hais, K. (1975) *Anglická mluvnice.* Praha: SPN.

Harris, J. [1751] (1765) *Hermes, or a Philosophical Inquiry Concerning Universal Grammar.* 2d ed. London: H. Woodfall.

Hopper, P. J. and E. C. Traugott (1993) *Grammaticalization.* Cambridge: Cambridge University Press.

Howe, S. (1996) *The Personal Pronouns in the Germanic Languages.* Berlin/New York: de Gruyter.

Hrnčíř, A. (1997) "Rozlišovanie rodu zvieracích podstatných mien v angličtine." *Acta Universitatis Matthiae Belii, Zborník Fakulty humanitných vied, Sekcia fiologická č. 1*, 159–168. Banská Bystrica: Univerzita Mateja Bela.

Huddleston, R.D. (1988) *Introduction to the Grammar of English.* Cambridge: Cambridge University Press.

Huddleston, R.D. and G.K. Pullum (2002) *The Cambridge Grammar of the English Language.* Cambridge: Cambridge University Press.

Huddleston, R. (1988) *English Grammar: An Outline.* Cambridge: Cambridge University Press.

Jolly, A. (1975) "Toward a Theory of Gender in Modern English." In *Studies in English grammar,* eds. A. Jolly and T. Fraser, 227–287. Paris: Editions universitaires.

Jones, C. (1988) *Grammatical Gender in English: 950–1250.* London: Croom Helm.

Jones, M. and P. Dillon (1987) *Dialect in Wiltshire.* Trowbridge: Wiltshire Country Council.

Kastovsky, D. (2000) "Inflectional Classes, Morphological Restructurring, and the Dissolution of Old English Grammatical Gender." In *Gender in Grammar and Cognition,* eds. B. Unterbeck et al., 709–727. Berlin: de Gruyter.

Kastovsky, D. and A. Mettinger (2001) *Language Contact in the History of English.* Frankfurt a.M.: Peter Lang.

Key, M.R. (1975) *Male/ Female Language*. New Yersey: Metuchen.

Lakoff, R. (1975) *Language and Woman's Place.* New York: Harper and Row.

Lakoff, R. (1990) *Talking Power.* New York: Basic Books.

Leech, G. and J. Svartvik (1975) *A Communicative Grammar of English.* London: Longman.

Mackay, D. G. and T. Konishi (1980) "Personification and the Pronoun Problem." *Women's Studies International Quarterly* 3: 149–163.

Marcoux, D. R. (1973) "Deviation in English Gender." *American Speech* 48: 98–107.

Mathiot, M. and M. Roberts (1979) "Sex Roles as Revealed Through Referential Gender in American English." In *Ethnolinguistics: Boas, Sapir and Wholf Revisited,* ed. M. Mathiot, 1–47. The Hague: Mouton.

Miko, F. (1962) *Rod, číslo a pád podstatných mien.* Bratislava: Vydavateľstvo SAV.

Mitchell, B. (1985) *Old English Syntax. Concord, the Parts of Speech, and the Sentence.* Oxford: Clarendon.

Moore, S. (1921) "Grammatical and Natural Gender in Middle English." *PMLA* 36: 79–103.

Morris, L. (1991) *Gender in Modern English: The System and its Uses*. PhD. thesis, Quebec: Université Laval.

Murphy, R. (1990) *English Grammar in Use*. Cambridge: Cambridge University Press.

Murray, J., Bradley, H. and W. Craigie (1994) *Oxford English Dictionary*. On CD-ROM (OUP) 2nd edition. Oxford.

Murray, L. [1975] 1968. *English Grammar*. Reprint. Menston: Scholar.

Pauwels, A. (1998) *Women Changing Language*. London and New York: Longman.

Pinker, S. (1994) *The Language Instinct*. New York: Harper Collins Publishers.

Porreca, K.L. (1984) "Sexism in Current ESL Textbooks." *TESOL Quarterly* 18: 705–724.

Prčíková, M. (1999) *"Gender and Animals."* In Acta Facultatis Universitatis Prešoviensis. Společenskovedný zošit 18, 85–141. Prešov.

Priestley, J. (1761) *The Rudiments of English Grammar*. London: R. Griffiths.

Quirk, R. (1995) *Grammatical and Lexical Variance in English*. London: Longman.

Quirk, R. and S. Greenbaum (1973) *A University Grammar of English*. London: Longman.

Quirk, R. Greenbaum, S., Leech, G. and J. Svartvik (1972) *A Grammar of Contemporary English*. London: Longman.

Quirk, R., Greenbaum, S., Leech, G. and J. Svartvik (1985) *A Comprehensive Grammar of the English Language*. London: Longman.

Quirk, R. and S. Greenbaum (1973) *A University Grammar of English*. London: Longman.

Schrampfer A. B. (1989) *Understanding and Using English Grammar*. NewYork: Prentice Hall.

Simon, S. (1996) *Gender in Translation*. London: Routledge.

Spender, D. (1980) *Man Made Language*. London: Routledge and Kegan Paul.

Sunderland, J. (1994) *Exploring Gender. Questions and Implications for English Language Education*. New York: Prentice Hall.

Swan, M. and C. Walter (1997) *How English Works*. Oxford: Oxford University Press.

Tannen, D. (1990) *You Just Don't Understand. Women and Men in Conversation*. New York: Ballentine Books.

Tannen, D. (1986) *That's Not What I Meant! How Conversational Style Makes or Breaks Your Relations with Others*. New York: William Morrow.

Tannen, D. (1990) *Women and Men in Conversation*. NewYork: Ballentine Books.

Tannen, D. (1993) *Gender and Conversational Interaction.* Oxford: Oxford University Press.

Tannen, D. (1994) *Gender and Discourse.* Oxford: Oxford University Press.

Tannen, D. (1996) "Gender in Research of Language." *TESOL Quaterly* 30: 341–344.

Thorne, B., Kramarae, C. and N. Henley (1983) *Language, Gender and Society.* Rowley, Mass.: Newbury House.

Vachek, J. (1964) "Notes on Gender in Modern English." *Sborník prací filosofické fakulty brněnské university* A 12: 189–194.

Valdrová, J. (1997) "K české genderové lingvistice." *Naše řeč* 80: 87.–91.

Visser, F. T. (1963) *An Historical Syntax of the English Language. Part I: Syntactical Units with One Verb.* Leiden: Brill.

Vomlela, J. (2003) "Gender and Animal Nouns in Selected Web Pages Devoted to Animal Keeping and Breeding." In *Theory and Practice in English Studies.* Volume I. Proceedings from the Seventh Conference of English, American and Canadian Studies, ed. J. Chovanec, 137–142. Brno: Masarykova univerzita.

Wagner, S. (2003) *Gender in English Pronouns - Myth and Reality.* PhD thesis, Universität Freiburg.

Wagner, S. (2000) "Depends How Long you Want for it to Take. *For/to* Clauses in Present-day Spoken British English." *Arbeiten aus Anglistik und Amerikanistik* 25: 191–211. Universität Freiburg.

Wagner, S. (2004) "'Gendered' Pronouns in English Dialects – A typological Perspective." In *Dialectology Meets Typology,* ed. B. Kortmann, 479–496. Berlin/New York: de Gruyter.

Willis, D. (1991) *Collins Cobuild Students' Grammar. Practice Material.* London: Harper Collins.

DUISBURGER ARBEITEN ZUR SPRACH- UND KULTURWISSENSCHAFT
DUISBURG PAPERS ON RESEARCH IN LANGUAGE AND CULTURE

Herausgegeben von
Ulrich Ammon, René Dirven und Martin Pütz

Band 1 René Dirven / Yvan Putseys (Eds.): A User's Grammar of English: Word, Sentence, Text, Interaction - Part A. The Structure of Words and Phrases. 1989.

Band 2 René Dirven / Richard A. Geiger (Eds.): A User's Grammar of English: Word, Sentence, Text, Interaction - Part B. The Structure of Sentences. 1989.

Band 3 René Dirven / Wolfgang Zydatiß / Willis J. Edmondson (Eds.): A User's Grammar of English: Word, Sentence, Text, Interaction - Parts C & D. The Structure of Texts / The Structure of Interaction. 1989.

Band 4 René Dirven: (Ed.): A User's Grammar of English: Word, Sentence, Text, Interaction - Compact Edition. 1989.

Band 5 Martin Pütz / René Dirven (Eds.): Wheels within Wheels. Papers of the Duisburg Symposium on Pidgin and Creole Languages. 1989.

Band 6 Ulrich Schmitz / Rüdiger Schütz / Andreas Kunz (Eds.): Linguistic Approaches to Artificial Intelligence. 1990.

Band 7 Friedrich Lenz: Organisationsprinzipien in mündlicher Fachkommunikation. 1989.

Band 8 Luc Van Doorslaer (Hrsg.): Niederländische Literatur im Spiegel niederländischer Kultur. Aufsätze zur Gastprofessur von Prof. Dr. Marcel Janssens (Universität Löwen) in Duisburg, Sommersemester 1988. 1990.

Band 9 Frank Lüschow / Marita Pabst-Weinschenk (Hrsg.): Mündliche Kommunikation als kooperativer Prozeß. Sprechwissenschaftliche Arbeitsfelder. Festschrift für Elmar Bartsch. 1991.

Band 10 Susanne Niemeier: Ein Fall im Medienvergleich: Film- und Fernsehversion von "A Streetcar Named Desire". Beitrag zur Annäherung an eine Mediensemiotik. 1990.

Band 11 Paul Kent Andersen: A new look at the passive. 1991.

Band 12 Richard Kremer: The efficiency of English language teaching in smaller countries: Denmark. Basic English language teaching in Denmark and its wider context. 1991.

Band 13 Frank Lüschow: Sprache und Kommunikation in der technischen Arbeit. 1992.

Band 14 Günter Kellermann / Michael D. Morrissey (Eds.): Diachrony within Synchrony: Language History and Cognition. Papers from the International Symposium at the University of Duisburg 26-28 March 1990. 1992.

Band 15 Wim Waumans (Hrsg.): Mit fremden Augen. Niederländisch: eine Sprache, verschiedene Kulturgemeinschaften. 1993.

Band 16 Eberhard Klein: Konditionalität in didaktischen und pädagogischen Grammatiken des Englischen. Prinzipien zur Erstellung von Lehr- und Lerngrammatiken. 1993.

Band 17 Xiaoan Zhu: "Wenn sich das Gras bewegt, dann muß auch der Wind blasen!" Studien zur Metapher in der deutschen politischen Pressesprache – unter besonderer Berücksichtigung der China-Berichterstattung. 1993.

Band 18 Fritz Ponelis: The Development of Afrikaans. 1993.

Band 19 Martin Pütz: Sprachökologie und Sprachwechsel. Die deutsch-australische Sprechgemeinschaft in Canberra. 1994.

Band 20 Heiner Pürschel (Ed.); Co-Editors: Elmar Bartsch, Peter Franklin, Ulrich Schmitz, Sonja Vandermeeren: Intercultural Communication. Proceedings of the 17th International L.A.U.D. Symposium. Duisburg, 23-27 March 1992. 1994.

Band 21 Dominiek Sandra: Morphology in the reader's mental lexicon. 1994.

Band 22 René Dirven: Metaphor and Nation. Metaphors Afrikaners live by. 1994.

Band 23 Ralph Bisschops: Die Metapher als Wertsetzung. Novalis, Ezechiel, Beckett. 1994.

Band 24 René Dirven / Johan Vanparys (Eds.): Current Approaches to the Lexicon. A Selection of Papers Presented at the 18th LAUD Symposium, Duisburg, March 1993. 1995.

Band 25 Michael Schloßmacher: Die Amtssprachen in den Organen der Europäischen Gemeinschaft. Status und Funktion. 1996. 2., durchges. Aufl. 1997.

Band 26 Johan Vanparys: Categories and Complements of Illocutionary Verbs in a Cognitive Perspective. 1996.

Band 27 Hideaki Takahashi: Die richtige Aussprache des Deutschen in Deutschland, Österreich und der Schweiz nach Maßgabe der kodifizierten Normen. 1996.

Band 28 Angela Heidemann: The visualization of foreign language vocabulary in CALL. 1996.

Band 29 Kazuma Matoba: Referenzperspektive in Sprechakten. Ihre Funktion und Entwicklung in der deutschen und japanischen Sprache. 1997.

Band 30 Olaf Jäkel: Metaphern in abstrakten Diskurs-Domänen. Eine kognitiv-linguistische Untersuchung anhand der Bereiche Geistestätigkeit, Wirtschaft und Wissenschaft. 1997.

Band 31 Birgit Smieja / Meike Tasch (Eds.): Human Contact through Language and Linguistics. 1997.

Band 32 Martin Pütz (Ed.): The Cultural Context in Foreign Language Teaching. 1997.

Band 33 Andrea Gerbig: Lexical and Grammatical Variation in a Corpus. A Computer-Assisted Study of Discourse on the Environment. 1997.

Band 34 Helga Bister-Broosen: Sprachkontakte und Sprachattitüden Jugendlicher im Elsaß und in Baden. Vergleichende soziolinguistische Untersuchungen in Colmar (Frankreich) und in Freiburg und Müllheim (Deutschland). 1998.

Band 35 Helga Bister-Broosen (Hrsg.): Niederländisch am Niederrhein. 1998.

Band 36 Alicja Sakaguchi: Interlinguistik. Gegenstand, Ziele, Aufgaben, Methoden. 1998.

Band 37 Benjamin Biebuyck / René Dirven / John Ries (Eds.): Faith and Fiction. Interdisciplinary Studies on the Interplay between Metaphor and Religion. A Selection of Papers from the 25th LAUD-Symposium of the Gerhard Mercator University of Duisburg on 'Metaphor and Religion'. 1998.

Band 38 Ulrich Ammon (Hrsg.): Sprachförderung. Schlüssel auswärtiger Kulturpolitik. 2000.

Band 39 El Hadj Ibrahima Diop: Das Selbstverständnis von Germanistikstudium und Deutschunterricht im frankophonen Afrika. Vom kolonialen Unterrichtsfach zu eigenständigen Deutschlandstudien und zum praxisbezogenen Lernen. 2000.

Band 40 Herman M. Batibo / Birgit Smieja (Eds.): Botswana: The Future of the Minority Languages. 2000.

Band 41 Harald Haarmann: Die Kleinsprachen der Welt – Existenzbedrohung und Überlebenschancen. Eine umfassende Dokumentation. 2001.

Band 42 Dirk Scholten: Sprachverbreitungspolitik des nationalsozialistischen Deutschlands. 2000.

Band 43 Riana Roos Paola: Pro-active Language Teacher Education in a Multicultural Society. 2001.

Band 44 Angelika Daniel: Lernerwortschatz und Wortschatzlernen im bilingualen Unterricht. 2001.

Band 45 Edel O'Halloran: Ist Mode englisch? Französische und englische Einflüsse auf die deutsche Mode- und Gemeinsprache im 20. Jahrhundert. 2001.

Band 46 Maya Khemlani David (Ed.): Methodological and Analytical Issues in Language Maintenance and Language Shift Studies. 2002.

Band 47 Martin Praxenthaler: Die Sprachverbreitungspolitik der DDR. Die deutsche Sprache als Mittel sozialistischer auswärtiger Kulturpolitik. 2002.

Band 48 Ulrich Ammon / Grant McConnell: English as an Academic Language in Europe. A Survey of its Use in Teaching. 2002.

Band 49 Birgit Smieja: Language Pluralism in Botswana - Hope or Hurdle? A sociolinguistic survey on language use and language attitudes in Botswana with special reference to the status and use of English. 2003.

Band 50 Carlos Inchaurralde / Celia Florén (Eds.): Interaction and Cognition in Linguistics. 2003.

Band 51 Andreas Ross: Europäische Einheit in babylonischer Vielfalt. Die Reform des Sprachenregimes der Europäischen Union im Spannungsfeld von Demokratie und Effizienz. 2003.

Band 52 Michael Anton Böhm: Deutsch in Afrika. Die Stellung der deutschen Sprache in Afrika vor dem Hintergrund der bildungs- und sprachpolitischen Gegebenheiten sowie der deutschen Auswärtigen Kulturpolitik. 2003.

Band 53 Martin Pütz / JoAnne Neff-van Aertselear / Teun A. van Dijk: Communicating Ideologies. Multidisciplinary Perspectives on Language, Discourse, and Social Practice. 2004.

Band 54 Yuko Sugita: Gesprächserwartungen. Eine kontrastive Studie über die Gesprächsführung in deutschen und japanischen Telefonaten. 2004.

Band 55 Eithne B. Carlin: A Grammar of Trio. A Cariban Language of Suriname. 2004.

Band 56 Maya Khemlani David (Ed.): Teaching of English in Second and Foreign Language Settings. Focus on Malaysia. 2004.

Band 57 Rita Brdar-Szabó / Elisabeth Knipf-Komlósi (Hrsg.): Lexikalische Semantik, Phraseologie und Lexikographie. Abgründe und Brücken. Festgabe für Regina Hessky. 2004.

Band 58 Claudia Kunschak: Language Variation in Foreign Language Teaching. On the Road to Communicative Competence. 2004.

Band 59 Allan Turner: Translating Tolkien. Philological Elements in *The Lord of the Rings*. 2005.

Band 60 Markus Motz (Hrsg.): Englisch oder Deutsch in Internationalen Studiengängen? 2005.

Band 61 Birte Kellermeier-Rehbein: Areale Wortbildungsvarianten des Standarddeutschen. Beiuntersuchung zum *Variantenwörterbuch des Deutschen*. 2005.

Band 62 Maya Khemlani David / Hafriza Burhanudeen / Ain Nadzimah Abdullah (Eds.): The Power of Language and the Media. 2006.

Band 63 Vít Dovalil: Sprachnormenwandel im geschriebenen Deutsch an der Schwelle zum 21. Jahrhundert. Die Entwicklung in ausgesuchten Bereichen der Grammatik. 2006.

Band 64 Sara Hägi: Nationale Varietäten im Unterricht Deutsch als Fremdsprache. 2006.

Band 65 Albert Weideman / Birgit Smieja (eds.): Empowerment through Language and Education. Cases and Case Studies from North America, Europe, Africa and Japan. 2006.

Band 66 Iraide Ibarretxe-Antuñano / Carlos Inchaurralde / Jesús-M. Sánchez-García (eds.): Language, Mind, and the Lexicon. 2007.

Band 67 Roswitha Reinbothe: Deutsch als internationale Wissenschaftssprache und der Boykott nach dem Ersten Weltkrieg. 2006.

Band 68 Fredericka van der Lubbe: Martin Aedler and the *High Dutch Minerva*. The First German Grammar for the English. 2007.

Band 69 Frank Polzenhagen: Cultural Conceptualisations in West African English. A Cognitive-Linguistic Approach. 2007.

Band 70 Augustin Simo Bobda (ed.): Explorations into Language Use in Africa. 2008.

Band 71 Johannes Eckerth / Sabine Siekmann (eds.): Task-Based Language Learning and Teaching. Theoretical, Methodological, and Pedagogical Perspectives. 2008.

Band 72 Laura Sicola: *No, they won't 'just sound like each other'*. NNS-NNS Negotiated Interaction and Attention to Phonological Form on Targeted L2 Pronunciation Tasks. 2009.

Band 73 Gerhard Stickel (ed.): National and European Language Policies. Contributions to the Annual Conference 2007 of EFNIL in Riga. 2009.

Band 74 Daniel Klaaß: Untersuchungen zu ausgewählten Aspekten des Konsonantismus bei österreichischen Nachrichtensprechern. 2009.

Band 75 Javier Herrero Ruiz: Understanding Tropes. At the Crossroads between Pragmatics and Cognition. 2009.

Band 76 Javier Valenzuela / Ana Rojo / Cristina Soriano (eds.): Trends in Cognitive Linguistics. Theoretical and Applied Models. 2009.

Band 77 Maya Khemlani David / James McLellan / Shameem Rafik-Galea / Ain Nadzimah Abdullah (eds.): Code Switching in Malaysia. 2009.

Band 78 Gerhard Stickel (ed.): Language Use in Business and Commerce in Europe. Contributions to the Annual Conference 2008 of EFNIL in Lisbon. 2010.

Band 79 Geert Brône: Bedeutungskonstitution in verbalem Humor. Ein kognitiv-linguistischer und diskurssemantischer Ansatz. 2010.

Band 80 Olga Martynova: Sprachwahl in der deutsch-russischen Unternehmenskommunikation. 2010.

Band 81 Gerhard Stickel (ed.): National, Regional and Minority Languages in Europe. Contributions to the Annual Conference 2009 of EFNIL in Dublin. 2011.

Band 82 Harald Haarmann: Writing as Technology and Cultural Ecology. Explorations of the Human Mind at the Dawn of History. 2011.

Band 83 Beate Höhmann: Sprachplanung und Spracherhalt innerhalb einer pommerischen Sprachgemeinschaft. Eine soziolinguistische Studie in Espírito Santo / Brasilien. 2011.

Band 84 Gabriele Schmidt: Motives for Studying German in Australia. Re-examining the Profile and Motivation of German Studies Students in Australian Universities. 2011.

Band 85 Michael Szurawitzki: Der thematische Einstieg. Eine diachrone und kontrastive Studie auf der Basis deutscher und finnischer linguistischer Zeitschriftenartikel. 2011.

Band 86 Azirah Hashim / Maya Khemlani David / James McLellan: Text, Discourse and Society. Functional and Pragmatic Approaches to Language in Use. 2011.

Band 87 Gerhard Stickel / Tamás Váradi (eds.): Language, Languages and New Technologies. ICT in the Service of Languages. Contributions to the Annual Conference 2010 of EFNIL in Thessaloniki. 2011.

Band 88 Monika Reif: Making Progress Simpler? Applying Cognitive Grammar to Tense-Aspect Teaching. 2012.

Band 89 Jan Kruse: Das Barcelona-Prinzip. Die Dreisprachigkeit aller Europäer als sprachenpolitisches Ziel der EU. 2012.

Band 90 Maria-Antonia Kontostathi: Die Arbeit mit dem Europäischen Sprachenportfolio als Möglichkeit zum selbständigen Erarbeiten fremdsprachlicher Kompetenzen. 2012.

Band 91 Yu Chen: Verbessern chinesische Studierende ihre Sprechfertigkeit im Deutschen während des Fachstudiums in Deutschland? Eine empirische Untersuchung unter Berücksichtigung sozialer Aspekte. 2012.

Band 92 Jakob Haselhuber: Mehrsprachigkeit in der Europäischen Union. Eine Analyse der EU-Sprachenpolitik, mit besonderem Fokus auf Deutschland. Umfassende Dokumentation und Perspektiven für die Zukunft. 2012.

Band 93 Hans Wagener: Untergräbt Deutschland selbst die internationale Stellung der deutschen Sprache? Eine Folge der Förderung von Englisch im Bildungsbereich. 2012.

Band 94 G. Stickel / M. Carrier (eds.) · Language Education in Creating a Multilingual Europe. Contributions to the Annual Conference 2011 of EFNIL in London. 2012.

Band 95 Jun He: Die Auswirkungen der englischsprachigen Hochschullehre in Deutschland auf das Deutschlernen in China. 2013.

Band 96 Monika Reif / Justyna A. Robinson / Martin Pütz (eds.): Variation in Language and Language Use. Linguistic, Socio-Cultural and Cognitive Perspectives. 2013.

Band 97 Carla Figueira: Languages at War. External Language Spread Policies in Lusophone Africa. Mozambique and Guinea-Bissau at the Turn of the 21st Century. 2013.

Band 98 Siew Imm Tan: Malaysian English. Language Contact and Change. 2013.

Band 99 Gerhard Stickel / Tamás Váradi (eds.): Lexical Challenges in a Multilingual Europe. Contributions to the Annual Conference 2012 of EFNIL in Budapest. 2013.

Band 100 Liudmila Li: Die Problematik der sprachlichen Integration von ImmigrantInnen. Unter Berücksichtigung des staatlich geforderten Sprachniveaus B1 (GER). Verbesserungsvorschläge auf Grundlage der empirischen Untersuchung in Berlin-Moabit. 2014.

Band 101 Xiaoyan Xia: Categorization and L2 Vocabulary Learning. A Cognitive Linguistic Perspective. 2014.

Band 102 Gerhard Stickel (ed.): Translation and Interpretation in Europe. Contributions to the Annual Conference 2013 of EFNIL in Vilnius. 2014.

Band 103 Maciej Mackiewicz: Interkulturelle Motivation im Fremdsprachenunterricht. Eine komparative Studie zu Deutsch als Fremdsprache in Polen und den USA. 2014.

Band 104 Silke Gester / Erika Kegyes (Hrsg.): Quo vadis, DaF? II. Betrachtungen zu Deutsch als Fremdsprache in den Ländern der Visegrád-Gruppe. 2015.

Band 105 Miroslava Majtanova: Die Rolle der deutschen Sprache für die Gruppenidentität von Deutschen im Ausland. Am Beispiel des Vereinslebens in Kuala Lumpur. 2015.

Band 106 Julia Wesian: Danksagungen in Dissertationen. Zur Genese einer Textsorte. 2015.

Band 107 Hector Kamdem Fonkoua: A Dictionary of Camfranglais. 2015.

Band 108 Frauke Priegnitz: Zwischen englischsprachigem Studium und landessprachigem Umfeld. Internationale Absolventen deutscher und dänischer Hochschulen. 2015.

Band 109 Gerhard Stickel / Cecilia Robustelli (eds.): Language Use in University Teaching and Research. Contributions to the Annual Conference 2014 of EFNIL in Florence. 2015.

Band 110 Wenting Sheng: Sprachförderungspolitik Deutschlands, Großbritanniens und Chinas im Vergleich. 2015.

Band 111 Eglė Kontutytė/Vaiva Žeimantienė (Hrsg.): Sprache in der Wissenschaft. Germanistische Einblicke. 2016.

Band 112 Ludmila Zemková: The Use of Gender Markers in Animals. As Demonstrated by Issues of National Geographic. 2015.

www.peterlang.com

www.ingramcontent.com/pod-product-compliance
Ingram Content Group UK Ltd.
Pitfield, Milton Keynes, MK11 3LW, UK
UKHW041826031026
15587UKWH00002B/602

9 783631 674574